Integrated Urban Water Management: Humid Tropics

T0186563

Urban Water Series – UNESCO-IHP

ISSN 1749-0790

Series Editors:

Čedo Maksimović
Department of Civil and Environmental Engineering
Imperial College
London, United Kingdom

J. Alberto Tejada-Guibert
International Hydrological Programme (IHP)
United Nations Educational, Scientific and Cultural Organization (UNESCO)
Paris, France

Sarantuyaa Zandaryaa
International Hydrological Programme (IHP)
United Nations Educational, Scientific and Cultural Organization (UNESCO)
Paris, France

Integrated Urban Water Management: Humid Tropics

Edited by

Jonathan N. Parkinson, Joel A. Goldenfum, and Carlos E.M. Tucci

UNESCO
Publishing

United Nations
Educational, Scientific and
Cultural Organization

CRC Press
Taylor & Francis Group
Boca Raton   London   New York   Leiden

CRC Press is an imprint of the
Taylor & Francis Group, an **informa** business

A BALKEMA BOOK

Cover illustration

Flood in Vila Nova district in Esteio, Brazil, 2007 – Courtesy of Charles Scholl Carvalho

Published jointly by

The United Nations Educational, Scientific and Cultural Organization (UNESCO)
7, place de Fontenoy
75007 Paris, France
www.unesco.org/publishing

and

Taylor & Francis The Netherlands
P.O. Box 447
2300 AK Leiden, The Netherlands
www.taylorandfrancis.com – www.balkema.nl – www.crcpress.com
Taylor & Francis is an imprint of the Taylor & Francis Group, an informa business, London, United Kingdom.

Typeset by MPS Ltd. (a Macmillan Company), Chennai, India
Printed and bound in Poland by Poligrafia Janusz Nowak, Poznán

ISBN UNESCO, paperback: 978-92-3-104065-8
ISBN Taylor & Francis, hardback: 978-0-415-45352-3
ISBN Taylor & Francis, paperback: 978-0-415-45353-0
ISBN Taylor & Francis e-book: 978-0-203-88117-0

Urban Water Series: ISSN 1749-0790

Volume 6

The designations employed and the presentation of material throughout this publication do not imply the
expression of any opinion whatsoever on the part of UNESCO or Taylor & Francis concerning the legal status
of any country, territory, city or area or of its authorities, or the delimitation of its frontiers or boundaries.
The authors are responsible for the choice and the presentation of the facts contained in this book and for the
opinions expressed therein, which are not necessarily those of UNESCO nor those of Taylor & Francis and do
not commit the Organization.

British Library Cataloguing in Publication Data
A catalogue record for this book is available from the British Library

Library of Congress Cataloging-in-Publication Data
Applied for

Foreword

Coping with too much water during most months of the year, and too little during the dry and hot pre-monsoon months, is a major problem in the humid tropics. The humid tropics – home to more than half the world's population, living mainly in developing countries – are characterized by frequent storms and high surface runoff. Yet urban water systems and infrastructures in most of the countries concerned are not capable of handling local rainfall intensity and duration. The key challenge in urban water management in these regions is, therefore, managing stormwater drainage and runoff pollution. Controlling water-related diseases related to sustained high temperatures and poor water supply and sanitation conditions are also common challenges in tropical countries. Furthermore, the impact of climate change and climate variability on urban water management appears more evident in the humid tropics.

In focusing on the specificities of the humid tropics, this book aims to contribute to the broader goal of sustainable urban water management across the world. It addresses a range of issues, including water supply, wastewater and stormwater, health hazards, and vulnerability to water-related disasters. Institutional, legal and socioeconomic issues, as well as education and capacity building, are also discussed, with examples and case studies.

This book presents the results of the "Humid Tropics" component of UNESCO's project on "Integrated Urban Water Management in Specific Climates", implemented during the Sixth Phase of UNESCO's International Hydrological Programme (2002–2007). The deliberations of an ad hoc workshop held in Iguazu Falls, Brazil, in 2005 and ensuing efforts by experts and collaborators have come together in the publication of this important book. The contribution of Carlos Tucci and Joel Goldenfum (both of the Federal University of Rio Grande do Sul, Brazil) and of Jonathan Parkinson (International Water Association, UK) as the editors of the book, leading an international team of experts in its preparation, is gratefully acknowledged. The publication, which is part of UNESCO-IHP Urban Water Series, was prepared under the responsibility and coordination of J. Alberto Tejada-Guibert, Deputy-Secretary of IHP and Responsible Officer for the Urban Water Management Programme of IHP, and Sarantuyaa Zandaryaa, Programme Specialist in urban water management and water quality at UNESCO-IHP. The role of Čedo Maksimović (Imperial College, London) in shaping the concept of the series on urban water management in specific climates is likewise acknowledged with appreciation.

UNESCO extends its gratitude to all the contributors for their outstanding effort, and is confident that the conclusions and recommendations presented in this volume will be of value to urban water management practitioners, policy-makers and educators alike in tropical regions throughout the world.

International Hydrological Programme (IHP)
United Nations Educational, Scientific and Cultural Organization (UNESCO)

Table of Contents

List of Figures

List of Tables

List of Boxes

Acronyms

BOD	Biochemical Oxygen Demand
CDC	Centers for Disease Control and Prevention
DAF	Dissolved Air Flotation
DALY	Disability Adjusted Life Year
ECLA	Economic Commission for Latin America
EPA	Environmental Protection Agency
GIS	Geographical Information System
IBAM	Instituto Brasileiro de Administração Municipal (Brazilian Institute of Municipal Administration)
IFRC	International Federation of Red Cross and Red Crescent Societies
IPCC	Intergovernmental Panel on Climate Change
ISDR	International Strategy for Disaster Reduction
IUWM	Integrated Urban Water Management
IUWP	Integrated Urban Water Plan
IWRM	Integrated Water Resource Management
LID	Low Impact Development
MDG	Millennium Development Goal
NFIP	National Flood Insurance Program
OSD	On-site Detention
RSF	Rapid Sand Filtration
SODIS	Solar Disinfection
SSF	Slow Sand Filtration
THMs	Trihalomethanes
UASB	Upflow Anaerobic Sludge Blanket reactor
UNDP	United Nations Development Programme
UNEP	United Nations Environment Programme
UNESCO	United Nations Educational, Scientific and Cultural Organization
UNIDO	United Nations Industrial Development Organization
UV	Ultraviolet light
WHO	World Health Organization
WMO	World Meteorological Organization
WSP	Waste Stabilization Ponds
WSP	Water Safety Plan
WSS	Water Supply and Sanitation

Glossary

Attenuation reduction of discharge and/or pollutant concentrations usually by storage either in a tank or within the urban drainage/sewerage system itself.

Attenuation tank a storage tank constructed to store flood flows and thus mitigate flooding.

Catchment the surface area bounded by topographical features, which drains to a single downstream location.

Combined sewer system a network of sewers which drains both foul wastewaters and stormwater runoff.

Combined Sewer Overflow an ancillary structure constructed as part of a combined sewer system which acts as a hydraulic valve when the sewers are surcharged.

Constructed wetland a bed colonized by macrophytic vegetation used for wastewater treatment.

Detention basin a storage pond constructed to store flood flows and thus mitigate flooding.

Detention time the average time that water is held in some storage.

Disability adjusted life year (DALY) the sum of the number of years of life lost due to premature mortality caused by a disease and the number of years lost due to disability as a result of affliction with the disease.

Disinfection a physical or chemical process that destroys pathogenic organisms.

Eutrophication process whereby nutrients from fertilizers or wastewaters elevate the nutrient concentrations and stimulate unnatural growths of algae or plants.

Excreta faeces and urine.

Flash flood a very high intensity flood which may occur at very short notice and cause considerable structural damage and damage.

Floodplain the flat land adjacent to river channels which is the first land to flood when the river levels rise.

Foul sewer a sewer which carries domestic sanitary sewage, commercial and industrial effluents.

Greywater wastewater that is produced as a result of personal washing, clothes washing, etc. (*see* Sullage).

Gross pollutants visible solids found in storm or dry weather flows which may include plastic and paper, vegetation, litter, excreta and discarded sanitary products.

Groundwater water that is stored in soils and voids between rocks beneath the ground surface.

Impermeabilization   covering of natural catchment surfaces with material (e.g. concrete, asphalt, tarmacadam) that does not allow the infiltration of water.

Infiltration   flow of water into the ground or into sewer pipe where the pipe is cracked or the joints are misaligned.

Microorganisms   unicellular living organisms so small that individually they can normally only be seen through a microscope, e.g. bacteria, *pathogens* and viruses.

Micropollutant   pollutants (organic matter, ammonium, bacteria) less than 45 microns (μm) diameter (e.g. inorganic pollutants and lead, zinc, cadmium etc.).

Non-point source pollution   a diffuse source of pollution which may result in chronic pollution.

On-site sanitation   On-site sanitation includes all forms of infrastructure for managing excreta that is located in the vicinity of the locality in which it is produced. On-site sanitation differs from on-plot sanitation, as it may not be located directly on the housing plot.

Overland flow   the runoff resulting from the surface depression storage being filled and the water flowing over the surface of the catchment towards a drainage system inlet.

Pathogens   microorganisms such as bacteria, viruses and protozoa, that cause sickness and disease in humans.

Point source pollution   a concentrated stream of waste resulting in localized pollution of the environment.

Receiving water   a natural body of water (e.g. river, lake or sea) into which waste is discharged.

Reuse   wastewater that is utilized for another purpose (e.g. stormwater runoff used for irrigation).

Sanitation   a system for promoting sanitary (healthy) conditions.

Septic tank   an underground storage and treatment device that is commonly used to treat domestic wastewater.

Sewage   a mixture of wastewaters from a variety of urban activities that utilize water.

Sewer   a closed conduit (usually a buried pipe) that is used to convey the wastewater from more than one property.

Sewerage   system of interconnected sewers comprising pipes buried in the ground to drain wastewaters.

Source control   technologies that are installed to promote local, on-site management and control of stormwater runoff close to the point of rainfall.

Stormwater   run-off caused by rainfall.

Sullage   wastewater from bathing, laundry, preparation of food, cooking and other personal and domestic activities that does not contain excreta. (Otherwise known as greywater.)

Swale   earth channel lined with grass into which runoff is drained and attenuated/treated.

Total solids   the total amount of solids that is generated in an urban catchment.

Trihalomethanes   chemical compounds in which three of the four hydrogen atoms of methane (CH4) are replaced by halogen atoms. Used extensively as solvents or refrigerants and are environmental pollutants, some of which are considered carcinogenic.

Urban stormwater management   the process of planning, designing, building, operating and restoring urban stormwater drainage systems (an inter-disciplinary subject involving several professional and trade skills).

Urbanization   the trend seen in many urban centres in which populations increase and density of inhabitation also increases.

Ventilated improved pit   a pit latrine with a screened vent pipe and a dark interior to the superstructure.

Virus   the smallest pathogenic microorganism that is capable of causing enteric disease.

Wastewater   all types of domestic wastewater (sewage or sullage), commercial and industrial effluent as well as stormwater runoff.

Water safety plan   a plan to ensure the safety of drinking water through the use of a comprehensive risk assessment and risk management approach that encompasses all steps in water supply from catchment to consumer.

Wastewater treatment   removal of pollutants or contaminants from sewage or wastewater for the protection of public health and the environment.

Watercourse   any stream or channel that carries or contains flowing water.

List of Contributors

Antônio Domingues Benetti
Institute of Hydraulic Research, Federal University of Rio Grande do Sul, Porto Alegre, Brazil

Luiza Cintra Campos
Department of Civil, Environmental and Geomatic Engineering, University College London, United Kingdom

Joel Avruch Goldenfum
Institute of Hydraulic Research, Federal University of Rio Grande do Sul, Porto Alegre, Brazil

Luis Eduardo Gregolin Grisotto
Information Centre in Environmental Health, University of São Paulo, São Paulo, Brazil

Jose Ochoa Iturbe
Catholic University of Andres Bello, Caracas, Venezuela

Tadeu Fabrício Malheiros
Engineering School of São Carlos, University of São Paulo, São Carlos, Brazil

Giuliano Marcon
Information Centre in Environmental Health, University of São Paulo, São Paulo, Brazil

Eduardo Mario Mendiondo
Engineering School of São Carlos, University of São Paulo, São Carlos, Brazil

Marllus Gustavo Ferreira Passos das Neves
Centre of Technology, Federal University of Alagoas, Maceió, Brazil

Jonathan Neil Parkinson
International Water Association, London, United Kingdom

Arlindo Philippi Jr
School of Public Health, University of São Paulo, São Paulo, Brazil

André Luiz Lopes da Silveira
Institute of Hydraulic Research, Federal University of Rio Grande do Sul, Porto Alegre, Brazil

Carlos Eduardo Morelli Tucci
Institute of Hydraulic Research, Federal University of Rio Grande do Sul, Porto Alegre, Brazil, and FEEVALE University Center, Novo Hamburgo, Brazil

Chapter 1

Integrated urban water management in the humid tropics

Carlos Eduardo Morelli Tucci

Institute of Hydraulic Research, Federal University of Rio Grande do Sul, Porto Alegre, Brazil
FEEVALE University Center, Novo Hamburgo, Brazil

This chapter presents the main issues related to Integrated Urban Water Management (IUWM) and their relevance within the humid tropics. Issues include urban development, climatic conditions, the various components of urban water services, and the impact of lack of these services on human and environmental health. The chapter also includes a discussion of the institutional aspects of IUWM, which act as the basis for implementation of these concepts in society.

1.1 CONCEPTS

Urbanization increases competition within a small space for the natural resources (air, land and water) required for human needs related to living, production and amenities. The natural environment and its human population together constitute a living and dynamic system that generates a set of interconnected effects, which, if not well-managed, can lead to a state of imbalance or even chaos.

The concept of sustainable development was developed to promote sustainable living in response to the economic and social pressures created by demands on natural resources by human society. Key components of a sustainable urban environment include environment conservation and the integration of health and socioeconomic aspects into urban development.

Integrated Urban Water Management

Urban water infrastructure services include water supply, wastewater and stormwater drainage facilities. Solid waste management is also closely related to urban wastewater management, particularly as solids have high pollutant loads associated with them, blocking drains and sewers and increasing flooding. Therefore solid waste management is viewed as an integral component of urban water management.

The main problems related to urbanization are predominantly caused by fragmented approaches to development and lack of capacity to effectively manage the rapid growth of unplanned settlements.

Urban master plans do not generally take into account all infrastructural components related to urban water management. As a result, outputs are poor with few indicators of efficiency.

Urban water facilities should:

- deliver potable water to the population (water supply)
- collect and treat the sewage produced by the city before its discharge into rivers or other receiving waters, thereby protecting the environment and its water sources (conservation for the future), and avoiding the spread of diseases (sanitation)
- install systems for stormwater drainage after urban occupation to mitigate any detrimental effects on the quality of receiving waters, and
- collect and dispose of solid wastes to reduce health risks from blockage and pollution of natural and engineered systems by solid wastes.

The main objectives of these services relate to security (stormwater flood control), health and environmental management.

Integrated Water Resource Management is increasingly being applied as a tool to promote sustainable management of water resources at the river catchment level. Cities generally either fall within a larger catchment or may comprise several small catchments. Cities often abstract water from upstream sources in a basin for supply and discharge effluents into downstream water bodies in the basin. These constitute external components of a city, which should be managed together with the main basin. Within the urban environment, Integrated Water Resources Management (IWRM) is referred to as *Integrated Urban Water Management (IUWM)*.

IUWM includes the management of water facilities and their interactions (Figure 1.1b) as part of Integrated Urban Management (Figure 1.1a). These components include: urban development (a driver based on economic and social development of the city), environment and health (the main goals), and institutional components, represented by a legal framework, management, capacity-building and monitoring of the required information for service management and development.

1.2 URBAN DEVELOPMENT IN THE HUMID TROPICS

The world is becoming increasingly urban, primarily as result of economic development and distribution of employment opportunities. In 1900, 13% of the global population was urban; by 2007 this had increased to almost 50%. However, this percentage occupied less than 3% of total global land area. In industrialized countries, population numbers are stable with a large urban population. But in developing countries, population numbers are still growing; by 2050 it is estimated that the world's population will have reached 9 billion, and most of this growth will occur in cities. By this time, urban population is forecasted to be nearly 70% of the world's population (UN, 2009).

In population terms, seventeen out of twenty of the largest cities in the world are in developing countries, most near or in the tropics. Large cities have grown enormously during the last century. For example, the metropolitan area of São Paulo in Brazil had

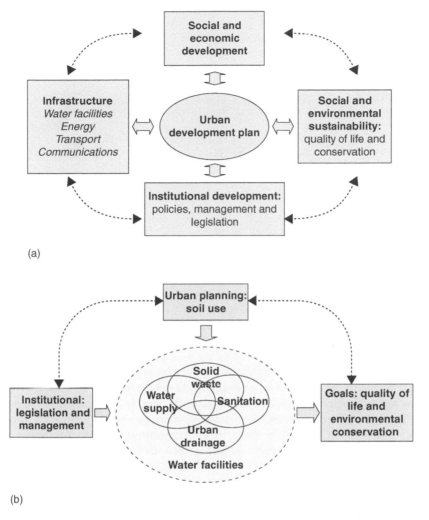

Figure 1.1 Urban management. a) Integrated urban management; b) Integrated Urban Water
    Management

Source: Tucci, 2009

about 200,000 inhabitants at the beginning of the twentieth century and 17 million at
the end of the century. This represents a thirty-four-fold increase with an average
yearly increase of 8.5%.

A significant part of the population lives in squatter settlements (*favelas* in Brazil or
barrios in Venezuela). In Caracas, more than 50% of the population live in this type of
area; in New Delhi, the figure is about 20%. Dwellings in slums are often built from
cardboard and scrap material, and are located in hazardous areas such as flooded
lands and steep hillsides. In the Amazon basin, in cities such as Belem and Manaus, a
large number of people live in *palafitas*, houses constructed on piles to raise the base
above the level of flood waters (see Figure 1.2).

Figure 1.2 *Palafitas* located on wetlands in Manaus, Amazon, Brazil

Source: Photo by C.E.M. Tucci

The main causes of the problems stated above are related to one or more of the following conditions:

- Poor people from rural areas migrate to towns and cities. For example, in 2004, Manaus, in the Amazon basin in Brazil, received about 40,000 economic migrants, attracted to the city by jobs. But these migrants often lack employment opportunities and live in substandard housing, which perpetuates a cycle of poverty and misery.
- Many cities do not have the capacity to accommodate this population increase, resulting in a large number of inhabitants in unregulated areas at the city border limits.
- City authorities and municipal managers have limited institutional capacity and lack adequate legislation and regulatory instruments.
- Urbanization is spontaneous but urban planning is conducted only for the portion of the city occupied by the middle and upper-income bracket. Slums are developed by an informal market without controls and demonstrate a lack of consideration for areas of risk, such as those subject to floods or mudslides. As a result, injury and death are commonplace during heavy storm events.
- Lack of Integrated Urban Water Management. The majority of water supply and sanitation services in cities do not take into account all the components of the urban water cycle. This results in the connection of sewers to the stormwater drainage system; a lack of, or inefficient wastewater treatment; flood increases due to lack of capacity in the stormwater system; excessive water losses in the water distribution network; and solids in the drainage system.

1.3 CLIMATIC CONDITIONS IN THE TROPICS

The tropics are those regions lying between the Tropic of Cancer and the Tropic of Capricorn (23°27′ N and 23°27′ S of the equator). The humid tropics are characterized

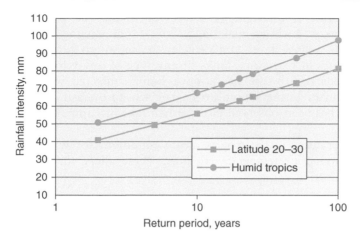

Figure 1.3 Comparison of the mean maximum rainfall of 1-hour duration gauges in the humid tropics and gauges inside the latitude of 20° and 30° S in Brazil

Source: Tucci and Porto, 2001

by a wet climate for more than seven months a year, an annual rainfall greater than 2000 mm, and a high mean temperature throughout the year.

The humid tropics contain parts of as many as sixty countries that are partly or completely located within this climate, including countries such as Brazil, Colombia, Costa Rica, Mexico, India, Myanmar, Thailand, Congo, Cameroon, Viet Nam, Malaysia and Indonesia. The National Research Council in the US estimated that the relative proportion of people living in the humid tropics accounts for about 45% of the population in the Americas, 30% in Africa, 25% in Asia, but only a small fraction in Oceania and the Pacific islands (NRC, 1993).

Many of the countries in the humid tropics can be classified as *developing*, based on socioeconomic indicators. Most water-related impacts can be traced to a combination of climatic conditions and the institutional and economic fragility of these countries.

In tropical climates, the temperatures and the intensity and frequency of rainfall are higher. Tropical and humid climates create many more difficulties for urban environmental management, stormwater constituting one of the main challenges. The principal climatic-related impacts on water and wastewater facilities in the humid tropics are as follows.

- Rainfall intensity is about 25% greater than in temperate climates. This requires more investment for the same level of drainage control, since peak discharges of runoff are higher in proportion to higher rainfall intensity (see Figure 1.3).
- Design conditions are based not only on local rainfall intensity, but also on rainfall duration. Low-intensity and long-duration storms maintain a high water level in the drains over long periods, creating a backing up of water from the large downstream (major) drainage system into the smaller, upstream (minor) drains, which contributes towads increased flooding. In this situation, the runoff exceeds the hydraulic capacity of the drainage system and streets are flooded. Examples of this situation can be seen annually during the monsoon season in cities in India and Bangladesh.

- High temperatures throughout the year encourage the development of water-related diseases such as malaria, yellow fever, dengue and schistosomiasis. For malaria, yellow fever and dengue, the mosquito host grows in wet and warm climates in reservoirs, tanks or any type of low-velocity water. The mosquito spreads the disease by biting a human infected with the disease then passing it on by biting a non-carrier. For schistosomiasis, the host develops in lakes. This has been observed in the urban lake of Pampulha in Belo Horizonte (Brazil).
- High temperatures in the tropics create conditions for water and waste treatment not found in temperate climates. These allow for the proliferation of microorganisms in waters.

1.4 WATER SUPPLY AND SANITATION

1.4.1 Water supply

The proportion of people worldwide with access to improved water supplies has risen from 77% in 1990 to 83% in 2002 (95% for the urban population). The population served with an improved water supply in developing countries increased by 8% between 1990 and 2002, which amounts to over 1,000 million more people served in twelve years (586 million in urban areas and 459 million in rural settings), representing 79% (WHO/UNICEF, 2002).

There are many vulnerable regions where about 460 million (8% of the total population) are already vulnerable or lack access to a potable and affordable water supply, while 25% of the total are moving towards a similar condition. Table 1.1 shows the proportion of populations in regions with access to improved water supply and sanitation systems. There are two main types of risk related to water supply:

Table 1.1 Proportion of urban population with improved[1] water supply and sanitation in 2006 (%)

Region	Water supply[2]	Sanitation[3]
Northern Africa	92	76
Sub-Saharan Africa	82	55
Latin America and Caribbean	92	79
Eastern Asia	90	65
Southern Asia	87	33
Western Asia	90	84
Sub-Saharan countries	58	31
Oceania	50	52
Developed regions	99	99
Developing regions	84	53

Note: 1. Improved water is a generic term referring to water delivered without population contamination. This is not the same as 'safe', which is based on specific indicators. 2. Water supply is understood as the water intended for use by the population. 3. Waste disposal is understood as the disposal of waste in a network or in the soil. It does not refer to the treatment of waste.

Source: WHO, 2009

- *Quantitative*: when there is insufficient water to supply the demand, in other words, inefficient water supply systems.
- *Qualitative*: when there is sufficient water volume, but the source of water is polluted and the water is therefore not potable. This scenario is common in developing countries, due to the lack of sewage treatment and the decreasing assimilative capacity of the rivers, followed by high urbanization rates.

The main indicators of water supply demand are: the population covered (as mentioned above), per capita consumption and losses from the water supply network. Some developed countries have high consumption rates. In developing countries, high demand is related mainly to the inefficiency of water distribution.

When urban communities are small, the solution for sewage disposal is local (for instance, the use of household or communal septic tanks, or local sewerage systems connected to small-scale wastewater treatment facilities). With population increases, sewage is collected by a centralized sewerage network, treated in large treatment plants and discharged into a river system.

The main difficulties encountered when moving from local solutions to network and treatment systems can be traced to the following factors:

- *Lack of political will*. Investment in these facilities is not usually a political priority. This reflects the lack of public pressure or environmental concern amongst the population. This scenario usually only occurs alongside concurrent economic development and a corresponding increase in income levels (Tietenberg, 2003).
- *Economic conditions*. Since these facilities are expensive, only medium to large cities possess the economic capacity to adopt these systems. However, cities are often dependent upon financing from central or regional governments.
- *Lack of integrated planning and institutional arrangement*. Sewer networks and treatment plants are designed and constructed, but not connected to houses. This situation arises either because the population sees no benefit in having a connection, or lacks the ability to pay for the service once connected. Since no planned institutional arrangement exists to enforce the system, the investment lacks cost recovery, and untreated sewage flows into the stormwater drainage system, and is directed into the rivers. As a result, the investment in sewage networks and treatment plants is lost.
- *Poor maintenance*. In this case, the system exists, but waste treatment efficiency is low. In combination with sewer and storm network interconnections, this results in negative impacts on the environment. Such problems are often related to a lack of goals defined by the institutions that manage the system.

1.5 STORMWATER AND FLOODPLAINS

Flood impacts on the population are due to:

- *Floodplains and hill slopes*. The population occupies risk areas such as floodplains and hill slopes during dry years, then floods cause problems in wet years.
- *Stormwater*. Urban development increases the extent of impervious areas, which increases peak flows and volume of runoff. Consequently, overland flows, flooding and impacts on the population increase.

Additional problems are related to the degradation of urban areas due to erosion and sedimentation, as well as water quality impacts from the wash-off of sediments and pollutants from urban surfaces.

1.5.1 Floodplain impacts

As the name suggests, flooding on floodplains is a natural event that occurs mainly in medium to large-sized river catchments. However, problems occur when human settlements form on these areas. The main impacts on populations occur as a result of lack of knowledge regarding the occurrence of flood levels and planning for space occupancy according to the risks of flood events.

A common scenario is uncontrolled urbanization on floodplains during a sequence of years with low flood levels. When high flood levels return, the damage increases and public administrations have to invest in flood relief. Structural solutions such as dams, dykes and river channel changes have high investment costs and only become feasible when damage costs are greater than development costs or as a result of intangible social factors. Non-structural measures have lower costs, but difficulties often arise with their implementation due to vested political interests and pressures.

1.5.2 Stormwater

Flooding is directly related to increases in impermeable areas and canalization, in which human-made conduits and drainage channels are designed to drain runoff as quickly as possible. These floods are not natural events and are created as a result of urban development.

Usually land surfaces in urban catchments are comprised of roofs, streets and other impervious surfaces. Runoff from these surfaces has a high velocity which adds to stormwater drainage problems. This increases peak flow and overland flow volume and decreases groundwater flow and evapotranspiration. Under these conditions, peak discharges increase together with frequency of flooding (Tucci and Porto, 2001; see Figure 1.4). Water quality deteriorates because the surface is washed during rainy days, increasing the pollution load in the urban environment and downstream rivers.

Cities that have developed according to conventional urban master plans do not generally consider the impact of urbanization on runoff and the associated problems with flooding. Neither the growth of impervious areas nor stormwater flows are regulated. City engineering departments do not have the hydrological support to cope with such problems through regulation, while engineering works – such as channels, pipes and installations – are designed without considering potential downstream impacts. Thus, flood frequency of drainage increases, with significant economic losses and, in some scenarios, loss of life.

1.5.3 Institutional issues of floodplain management

The main institutional issues of flood management are related to the occupation of flood risk areas on floodplains.

High-risk areas are frequently occupied by low-income communities or, during a succession of low annual peak flows, by populations with higher incomes who construct better housing and install better facilities. Poor populations usually migrate to

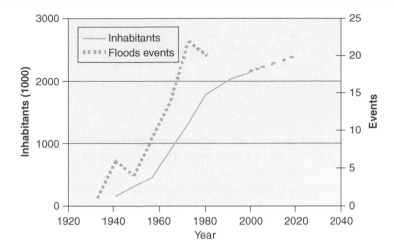

Figure 1.4 **Population increase and flood events in Belo Horizonte, Brazil**

Source: based on data from Ramos, 1998

these areas and receive support from the government to relocate, but others often take their place. However, this scenario tends not to occur when the area is developed with a basic infrastructure, public facilities and amenities.

The main issue is that urban planning does not take into account flood risk areas in relation to ongoing and future developments. For cities that have adopted flood zoning (for instance São Paulo and Curitiba), the main issue has been law enforcement in developments in both public and private areas. In public areas, there is the risk of invasion by low-income populations, whereas in private areas the issue relates primarily to illegal developments.

1.5.4 Stormwater management

The traditional model adopted for the management of urban runoff is based on a misconception, as it involves draining runoff from urban surfaces as quickly as possible through a system of pipes and channels. However, this increases peak flows and the costs of stormwater management. Moreover, as there is no control of peak increases at minor drainage levels, most of these impacts will affect the central drainage system.

To cope with this problem, many city and public administrations have developed additional works, such as channels. But, this type of solution only transfers flood problems from one section of the basin to another downstream, with high consequential costs. In addition, water quality decreases, since runoff contains a larger amount of solids and a higher concentration of metals and other toxic components.

Since the 1970s in developed countries, the concept of stormwater management has evolved through the application of new technologies, such as detention and retention ponds, permeable surfaces, infiltration trenches and other source control measures. This approach has been implemented through municipal regulations in developed countries and the cost of implementation and control is paid by the land developer.

However, in developing countries, this type of control usually does not exist and the impacts are transferred downstream into the major drainage system as described above. The cost of the control of this impact is transferred from the household to the public domain, since the municipality has to invest in hydraulic structures to reduce the downstream floods impacts.

1.6 TOTAL SOLIDS

The solids contained in stormwater include: *sediments* from the rainfall erosion of urban surfaces; *solid waste (rubbish)* such as plastic, papers and other refuse; and *vegetation* matter.

In the urban development basin scenario, two stages of solids production can be identified:

- At the initial stage of development, a large amount of sediment is present, compared to natural conditions, because of construction and loss of vegetal cover. Rainfall energy and increase in velocity of runoff from impervious areas increase soil erosion and transport more sediment into nearby urban creeks.
- After urbanization has stabilized, sediments remain as an important part of total solids. Neves (2006) showed that in a basin which is 67% urbanized, 77% of total solids come from sediments, stones and vegetation, and 23% constitute refuse. Solids waste increases mainly due to human activities – lack of efficient services and lack of education concerning street cleaning and waste collection.

In some Brazilian cities, sediment in major drainage varies from about 400 to 2450 $m^3 km^{-2}$ year as a result of different stages of urbanization. Assuming a density of 50 inhabitant/ha and a cost of US\$4.00/m^3, the yearly unit cost of maintenance per person is US\$0.32–1.96.

The total production of rubbish in Brazil was about 0.5 to 0.8 kg/(person per day). The mean value in Brazil in 2000 was 0.74 kg/(person per day) (IBGE, 2002). In San José, California, US, the refuse arriving in the drainage system was estimated at 1.8 kg/(person per year), but after street cleaning, there was a reduction of 0.8 kg/(person per year) of solids in the storm network (Larger et al., 1977). Armitage et al., (1998) stated that 3.34 m^3/ha/year of garbage is cleaned from streets in Springs, South Africa and 0.71 m^3/(ha per year) [82 kg/(ha per year)] arrives in the drainage.

In the basin of Porto Alegre, with an area of 1.92 km^2 (63% urbanized), Neves (2006) estimated that 1.95 kg/person/year of refuse is cleaned from the streets, 0.11 kg/person/year arrives in the drainage, and 0.33 kg/person/year is measured downstream of the basin. In this basin, 82% of the refuse is plastic and 50% is non-recyclable plastic. There is significant variability among these data because of other interrelated factors, such as efficiency of the service and education. The composition of the rubbish that arrives in the drainage varies according to type of urbanization, recycling and efficiency of the services in different neighbourhoods.

1.7 WATER QUALITY

Water quality is the result of urban occupation, sewage, stormwater and solids management. The main risks for the environment and population from these sources of

Box 1.1 Water supply contamination by eutrophication

When urbanization develops from downstream to upstream in the water supply basin without control of sewage and stormwater, nutrients from the development flow into the reservoir, changing its trophic conditions. This results in a eutrophic reservoir that allows algae to grow, which may produce toxicity in the water. In most cases, water treatment for domestic supply does not eliminate toxicity. This can have can accumulated consequences for human health over time and serious diseases may develop as a result.

Table 1.2 Comparison of mean values of water quality parameters from stormwater in several United States cities and Porto Alegre (Brazil) (mg/l)

Parameter	Durham (1)	Cincinnati (2)	Tulsa (3)	P. Alegre (4)	APWA (5) Interval	
					Lower	Upper
BOD		19	11.8	31.8	1	700
Total solids	1440		545	1523	450	14,600
Ph		7.5	7.4	7.2		
Coliforms (NMP/100 ml)	23,000		18,000	1.5×10^7	55	11.2×10^7
Iron	12			30.3		
Lead	0.46			0.19		
Ammonia		0.4		1.0		

(1) Colson (1974); (2) Weibel et al. (1964); (3) AVCO (1970); (4) Ide (1984); (5) APWA (1969)

pollution are: contamination of water supply sources such as rivers and aquifers, spread of disease, and modification of the fauna and flora of water systems (Box 1.1).

During dry weather periods, only sewage flows in the drainage system. On rainy days, the surface wash-load is mainly grit, litter and other solid waste from street surfaces. Most of the pollutant loads from stormwater occur in the first part of a storm event, as it is during this time that solids are washed off from urban surfaces (EPA, 1993).

The main difference in load characteristics between sewage and stormwater is that sewage is mainly organic with high concentrations of Biochemical Oxygen Demand (BOD) and bacteria (e.g. coliforms), while stormwater has higher concentrations of metals such as copper, zinc and lead. Table 1.2 shows some parameters of concentrations from stormwater in cities in the United States and Porto Alegre, Brazil.

1.8 WATER-RELATED DISEASES

Some of the main diseases related to water supply, sanitation and drainage in the humid tropics, are: diarrhoea, cholera, malaria, dengue and leptospirosis. More than fifty diseases are associated with poor sanitation, which results in millions of deaths, mainly of children (usually from diarrhoea). For example, Bangladesh has twice the number of infant deaths in urban slums than in urban areas as a whole (Wright, 1997).

Diarrhoea is a disease closely related to poor sanitation and hygienic behaviour, and is the main cause of child death in developing countries. An adequate water supply and good sanitation reduces child mortality by 55% (World Resources Institute, 1992).

Malaria is endemic in some countries, most of which are in humid tropic areas. Environmental conditions, related to drainage, that help to spread malaria are: stagnant waters, deforestation, soil erosion and flooding. Malaria is a disease strongly related to the humid tropics. A combination of environmental factors, such as forests (cloudiness), temperature and stagnant water, among others, are responsible for its development. The vector is a mosquito, genus *Anopheles*, whose proliferation is directly related to watercourses. Its ideal conditions are clean water, shade and small flow velocity. Urban malaria is an important issue and has been reported by Cgiarg (2009) and Kolsky (1999) with regard to Africa and Indian urban development. In the Brazilian humid tropics, there were nearly a million cases of malaria in 2000 and 2001 (Santos, 2005).

Dengue is a disease related to warm climates. Its spread depends on the *Aedes* mosquito, which lives in clean and stagnant water found in homes (tyres, vases, etc) during the rainy season. It has had a major impact in tropical cities such as Rio de Janeiro and Belo Horizonte in Brazil. On-site storage should be carefully designed in this type of climate to avoid creating an environment for this kind of disease.

Schistosomiasis uses the genus *Biomphalaria* as an intermediate host. *Biomphalaria* is the host of the larvae (miracidia) that develops in aquatic environments. Creeks, streams, lakes, wetlands and artificial water systems such as irrigation canals and small impoundments with some plants, favour the development of the snail (Santos, 2005).

The building of small dams, impoundments and other structures for storage and attenuation of runoff result in the proliferation of the intermediate host, snails, which act as the disease transmission vector in the tropics. The principal action taken has been to develop drainage that operates as fast as possible. However, this conflicts with certain strategies of temperate climates, which rely on storage as part of the process.

1.9 URBAN WATER – AN OVERVIEW OF THE MAIN ISSUES

In most developed countries, water supply, sewage treatment and stormwater flooding (due to urbanization) are no longer a major problem. As shown in Table 1.3, the main issues of concern relate to pollution from stormwater and the management of floodplain hazards (natural floods).

However, for many developing countries, access to basic sanitation remains the single most important issue and the investment requirements for the development of sanitation remain a major challenge. While total coverage may appear to be good, the majority of sanitation serves the wealthy areas of cities and the poor still lack access to basic facilities.

Sanitation coverage is one of the key indicators of urban poverty, since overcrowded and unhealthy living conditions for the urban poor in developing countries are made even more degrading by the lack of adequate systems to dispose of human waste (Wright, 1997). Table 1.3 compares developed and developing countries for each aspect of water in urban areas.

In addition, the disposal of untreated urban wastewater decreases the amount of clean water available for supply, and new investments have to be made to maintain

Table 1.3 Comparison of urban water management between developed and developing countries

Facility	Developed country	Developing country
Water supply	Total coverage with some risk of contamination from non-point sources	Lack of supply and contamination of water sources through lack of sanitation
Sanitation	High coverage of sewerage networks and sewage treatment	Low coverage and low efficiency of existing treatment and sewerage systems
Stormwater	*Quantitative control:* floods are regulated by a combination of structural and non-structural measures *water quality* is still an important issue that needs to be addressed.	1. Lack of measures for control of water quantity or quality with high level of impacts 2. Cost of the impacts are transferred to the public or to the environment 3. Poor investments which usually increase floods.
Flood hazard	Mainly non-structural measures with insurance, zoning and flood alerts.	1. Occupation of floodplain without control 2. Bad investments in structural solutions 3. Occupation of floodplains by the poor during drought season and high impact during flood season.

and improve water supplies. Most developing countries have adopted convenient planning 'layouts' in which the water supply intake is located upstream of the city enabling the use of clean water, while waste is discharged downstream without treatment for the river to dilute and assimilate the pollution load.

As a result of urbanization and consequent population increases, cities now cover large areas. In many cases, upstream cities discharge untreated or only partially treated wastewater into a river from which another downstream city abstracts water for its supply. This represents the *contamination cycle* of a river basin, as illustrated by Figure 1.5. This scenario results from the so-called *hygienist period* (Table 1.4) when significant benefits are attained from the introduction of basic sanitation, but excreta and pollutant impacts are transferred into sewage loads and stormwater.

Developed countries moved away from the *hygienist period* in the early 1970s and into the *correction* stage (see Table 1.4), which was characterized by high investment in sewerage networks and treatment plants that improve the quality of downstream waterbodies, together with regulation and stormwater management using detention and retention ponds to attenuate flows and reduces the downstream transfer of stormwater impacts.

After the 1990s, sustainable development concepts, such as *low impact developments* (LID), were introduced into the urban environment (NAHB, 2004). These developments were designed according to the concept of non-point source pollution from stormwater and the use of infiltration practices to restore the natural hydrological functions of catchments (Table 1.4). However, although some developing countries are trying to move into the second stage of development, many remain in the first stage, and all are a long way from the situation characterized by sustainable development.

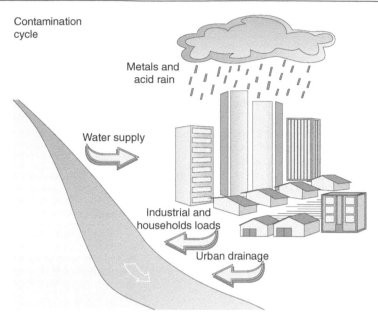

Figure 1.5 **Contamination cycle**
Source: Tucci, 2001

Table 1.4 **Stages of urban water developments in developed countries**

Stages	Period	Characteristics
Pre-hygienist	Until early twentieth century	Urban systems without sewer and stormwater networks and treatment; most households use some form of on-site sanitation; stormwater in the streets with high proliferation of water-related diseases.
Hygienist	Until 1970s	Potable water supply, sewer network without treatment plants causing river contamination; stormwater in channels, conduits or street flows, transferring pollutant impacts of runoff downstream.
Correction	After 1970s	Sewage treatment, detention and retention ponds in the stormwater systems to control peak flows of runoff from urban areas.
Sustainable	After 1990s	Introduction of better regulation and control measures to reduce stormwater pollution; drainage systems incorporate natural practices of infiltration and maintain natural hydrological functions. Urban planning takes into account stormwater runoff and the need to maintain natural flow conditions.

1.10 INTEGRATED URBAN WATER MANAGEMENT

1.10.1 Integrated aspects

Urban water management functions at two main levels:

- *The urban catchment.* This is the area of development inside the city.
- *The natural hydrological catchment.* The urban catchment lies within the natural hydrological catchment and the city abstracts water and returns it into the environment as waste. It is usually a major river basin where the municipality is an additional user.

Within the city boundaries, the management function rests at the municipal level, whereas at the basin level, management requires a state or national-level institution, usually in the form of a basin authority.

How these levels are connected in terms of institutional management is one of the key elements in IWRM and IUWM development. Table 1.5 presents a common scenario of integration of basin and city water planning at a national level.

1.10.2 IWRM 'outside the city'

The United Nations Millennium Development Goals (MDGs) give the improvement of water and sanitation in all countries as one of its main goals. In order to attain this goal, the UN Conference in Johannesburg recommended the implementation of National Water Plans as a main tool of IWRM.

Some countries, such as Brazil, France and Australia, already have national integrated water legislation. Other countries are expected to follow, as some are currently updating old legislation or are in the process of developing this form of legislation.

Table 1.5 Integration of management at basin and city levels

Space	Administration level	Management	Instrument	Characteristics
Basin[1]	Nation or State	Basin Committee and Agency	Basin Water Plan	Sustainable management of quantity and quality of the rivers in the basin avoiding impact transferences.[3]
Municipality[2]	or Metropolitan Area	Municipality	Integrated Urban Water Plan	Sustainable development of urban water facilities inside the city, avoiding transference of downstream impacts in the river system according to basin regulations.

1. Usually large basins ($>1000\,km^2$)
2. Area covered by the municipality and its small major drainage basins ($<50\,km^2$)
3. Impact transference: when urbanization takes place it could increase peak flood, flow velocity and sediment together with water quality pollution.

Table 1.6 Institutional development in Brazil

Phase	Period	Characteristics
Sectoral approach	1934–1997	Water fragmented by user sectors
National integrated water institutions	1997–2000	Establishment of National Legislation, National Water Council, National Water Agencies
Decentralization of government	2000–2006	States institutions, Basin committee and sectoral legislation
Sustainable development	2006–ongoing	Economic and political sustainability for water management and visible outputs to society and environment

Source: adapted from Tucci, 2006

Table 1.6 shows the steps and scenarios that took place in Brazil, which has been a model for institutional development, based on the use of the basin as the geographic space for management. However, the challenge remains of how to integrate 'urban' and 'basin' management functions.

In Chile, there has been considerable investment in increasing the level of wastewater treatment (Peña et al., 2004). However, in others countries, including Brazil, over the period 2001–2005, investment was below the population growth rate. This resulted in an overall decrease in the proportion of wastewater treated prior to discharge into the natural environment.[1] It has been estimated that Brazil requires about 0.6% of its GDP over twenty years to cover the overall needs (ESBE, 2006). This clearly indicates that if institutional arrangements are not followed by strategic investments it is not possible to expect marked and measurable outputs.

In order to achieve the MDGs related to water and sanitation, it is therefore important to understand the reality of each region or country and its needs. Usually the main needs are as follows:

● *A strong lack of knowledge and development of Integrated Urban Water Management.* Urban facilities are developed through a fragmented planning approach, where urban development and planning is disconnected from urban water systems. The water supply basin is already polluted because of a lack of wastewater treatment, decreasing the availability of water that can be used for potable water supplies without extensive treatment. A sewer network often exists, but without the requisite house connections to ensure that domestic wastewater discharges into it. An increase in impervious areas and channels exacerbates flood problems and pollutant loads from stormwater.[2]

---

[1] Brazilian deficit on sewage treatment is 71.8% (ESBE, 2006), which represents 7,000 to 11,000 of BOD load in the rivers every day.

[2] With help from international financing institutions, some Brazilian cities invested from US$25 to US$50 million/km to increase the flow capacity of channels, but this type of work increased peak flows downstream and exacerbated flooding.

- *Lack of institutional capacity and ineffective integration of national and local levels (municipalities), such as national basin committees and utilities responsible for water supply and sanitation (sewerage) at the municipal level.* When water plans and basin committees are developed, they should have a direct communication link to the urban water management structure. This has been a major problem in many countries, due to the difference in level of government (municipal, state and federal) and the different backgrounds of the professionals involved at the basin level and in WSS management.
- *Long-term permanent investments and a commitment to strategic planning to meet the goals.* The process of developing basin committees and many of the instruments of IWRM are important, but require a measurable output by society. For example, in the case of one of the oldest Brazilian basin committees, apart from discussion, there has been little real change in the way that water supplies and sanitation are managed for more than twenty years. During 2006, there were many occurrences of dying fish because of an oxygen deficit, due to the almost complete lack of wastewater treatment.

Concerning the achievement of the MDG goals in developing countries, there is a need for the following:

- A concrete programme within the general terms of water management. This should utilize the National Water Plan (which clearly incorporates the MDGs and the framework of institutional arrangement required), followed by a Strategic Action Plan, including the required source of funds. Based on these elements it should advise on the potential financing mechanisms to support the proposed actions in the country.
- The development of a follow-up process concerning measurable outputs. A Strategic Action Plan needs to be prepared at the country level and integrated with state and municipal levels. It should advise on international financing instruments, based upon sound institutional development and plans.

1.10.3 IUWM 'inside the city'

Overlapping urban services: Integrated Urban Water Management necessitates the development of planning and management components of urban water services: water and sanitation, stormwater and total solids. These services, together with urban development, have many overlapping areas. In most countries, these services have been developed by different institutions, which creates difficulties in managing common issues. As shown in Figure 1.6, the main overlapping issues in the systems are as follows:

- *Water supply*: (a) wastewater and stormwater discharges pollute the water supply basin and groundwater; (b) leachate from landfill sites pollutes groundwater and downstream rivers; (c) flood erosion may affect water supply and sanitation facilities.
- *Waste and stormwater*: (a) combined network systems for wastewater and stormwater are required by conventional design; (b) with separate systems many management difficulties occur when linking both systems.

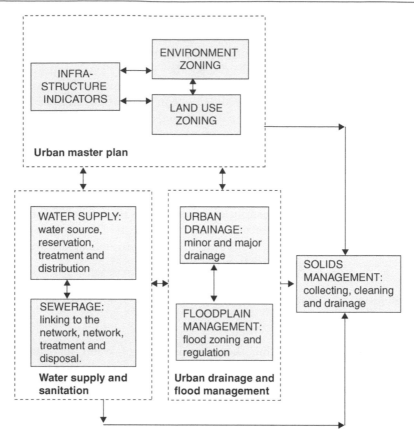

Figure 1.6 **Interconnection of urban water facilities (environment zoning is the selection of conservation and preservation areas in the urban area)**

Source: Tucci, 2001

- *Urban stormwater drainage and total solids*: (a) stormwater network efficiency is affected by street cleaning and solids waste collection services as described above; (b) drainage and soil erosion control require common strategies.

Institutional management of these facilities usually constitutes the central difficulty since:

- there are many institutions covering these services without an integrated management system
- some of the services are not measurable and public perception of the benefits tends to be low; as a result there is low feasibility for cost recovery of services for stormwater and solids management, and
- there is a lack of law enforcement and an effective regulatory agency. In this scenario, there are no indicators for services and pricing is usually high with low efficiency.

Management has been successful only when one institution manages all these services (for instance, Santo Andre in Brazil), together with a regulatory agency that enforces public rights and interests; in such cases, the population has a better perception of overall service. The fragmentation of urban facilities is the result of political arrangement, lack of knowledge and conservative technical thinking from decision-makers.

Implementation of Integrated Urban Water Management is developed in each scenario through the Integrated Urban Water Plan (IUWP). The plan could be developed using sub-plans for each component of the urban services, but with overlap in the reference terms used in order to achieve integration. Figure 1.7 presents the overall framework at the city level, showing the integration of plans and the institutional framework; Figure 1.8 presents the framework of the IUWP.

The Integrated Urban Water Plan (IUWP) or the specific plans form the main inputs into the Urban Development Master Plan and has the following components (Figure 1.9):

- *Conceptual*: definition of the conceptual aspects of the plan, such as, principles, strategies, scenarios covering the plan and the risks. Relevant definitions are required in the planning steps.

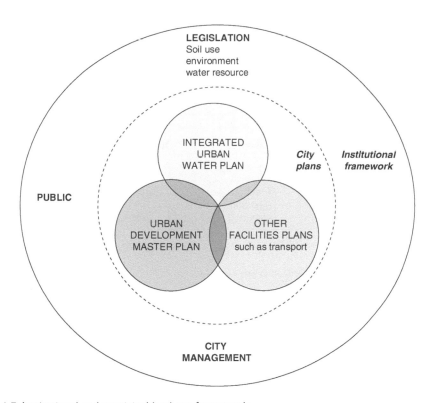

Figure 1.7 Institutional and municipal land use framework

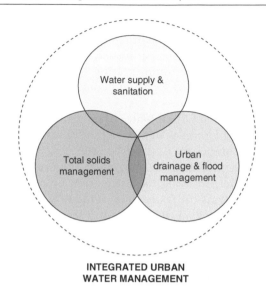

**INTEGRATED URBAN
WATER MANAGEMENT**

Figure 1.8 Components of integrated urban water management

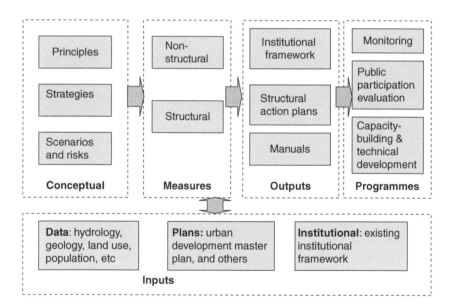

Figure 1.9 Characteristics of integrated urban water plans

Source: adapted from Tucci, 2001

- *Measures*: are based on the two main branches of activities:
  a) *Non-structural.* These are the legal and management developments in the city, which are strongly related to land use and population requirements for sustainable development. The legal aspects related to regulation and management develop the municipality structure for project analysis, implementation and enforcement. For example, the limits for stormwater impacts to downstream

areas can be regulated for new development by limiting the flow, sediment and water quality restrictions of municipality regulation. The ways in which this legislation is to be implemented under the municipality management have to be proposed in the IUWP.

b) *Structural*. These are the measures for implementing the urban facilities (water supply, sanitation, stormwater and total solids management). These are usually developed by the basins of sub-areas of the city and should take into account actual and future urban development forecasted by the city's Urban Development Master Plan.

- *Outputs*: include water regulation, taking into account all water facilities; the management of the municipality administration structure; water supply; sanitation; Stormwater and Total Solids Action Plans, including feasibility studies; and the time schedule for implementation.
- *Programmes*: are the long-term activities identified in the IUWP designed to complement the needs of the water facilities in the city over a long period. These may include activities such as monitoring of hydrologic and water quality variables, urban occupation and land use, impacts records and evaluations of these impacts.

1.10.4 Objectives and principles of urban stormwater and flood control master planning

The goals and objectives of urban planning are aimed at improving the well-being of the population and the protection of the environment. In stormwater and floodplain management the main goals are the following:

- Reduce flood risks by distribution volumes of runoff in time and space within the urban basin as a whole (deconcentration).
- Control the occupation of floodplain areas through regulation and other non-structural measures.
- Prevent and provide relief measures for low-frequency floods by flood warning.
- Improve stormwater water quality by developing on-site and stormwater measures.

The good experiences in flood control of many countries have now led to the development of a set of principles in stormwater management. These principles have been applied in developed countries, but are rarely fully utilized. Stormwater practices in most developing countries do not currently fulfil these principles, which are as follows:

- Flood control measures should take into account the whole basin and not only specific river reaches.
- Stormwater control scenarios should take account future city developments.
- Flood control measures should not transfer the flood impact to downstream reaches, giving priority to source control measures.
- The impact caused by urban surface runoff and other impacts related to stormwater water quality should be reduced.
- Management starts with the implementation of the Stormwater Master Plan in the municipality.
- Public participation in urban drainage management should be increased.

REFERENCES

APWA. 1969. Water pollution aspects of urban runoff. *Water Quality Administration.* Water Pollution Control Research Series. Report No. WP-20-15.

Armitage, N., Rooseboom, A., Nel, C. and Townshend, P. 1998. *The Removal of Urban Litter from Stormwater Conduits and Streams.* Report No. TT 95/98. Pretoria: Water Research Commission.

AVCO. 1970. Stormwater pollution from urban activity. *Water Quality Administration.* Water Pollution Control Research Series. Report No. 11034 FKL.

CGIARG. 2009. Malaria Knowledge Programme. Policy Brief. http://www.iwmi.cgiar.org/sima/files/pdf/Urban%20Malaria%20Policy%20Brief.pdf (accessed in 2009).

Colson, N.V. 1974. *Characterization and Treatment of Urban Land Runoff.* EPA. 670/2-74-096.

EPA. 1993. *Guidance Specifying Management Measures for Sources of Nonpoint Pollution in Coastal Waters.* Environmental Protection Agency. EPA 840-B-92-002. Washington DC. http://www.epa.gov/owow/nps/MMGI/index.html (accessed in 2006).

ESBE. 2006. PNAD 2005 Aumenta o déficit de dos serviços de Saneamento Básico. Projeções indicam Universalização em 50 anos. ESBE Associação das Empresas de Saneamento Estaduais.

IBGE. 2002. *Pesquisa Nacional de Saneamento Básico.* PNSB, 2000. http://www.ibge.net /ibge/presidência /noticias/27032002pnsb.shtm (accessed 27 March 2002).

Ide, C. 1984. *Qualidade da drenagem pluvial urbana.* Dissertação de mestrado, Programa de pós-graduação em Recursos Hídricos e Saneamento, IPH/UFRGS.

Kolsky, P. 1999. Engineers and urban malaria: part of the solution, or part of the problem? *Environment and Urbanization,* Vol. 11, No. 1, April.

Larger, J., Smith, W.G., Lynard, W.G., Finn, R.M. and Finnemore, E.J. 1977. *Urban Stormwater Management and Technology: Update and User's Guide.* US EPA Report 600/8-77-014 NTIS N. PB 275654.

NAHB Research Center. 2004. *Municipal Guide to Low Impact Development.* Maryland, US: NAHB. http://www.lowimpactdevelopment.org.

Neves, M. G. F. P. 2006. Quantificação de resíduos sólidos na drenagem urbana Tese de doutorado. Universidade Federal do Rio Grande do Sul, Porto Alegre, Brazil.

NRC. 1993. *Sustainable Agriculture and the Environment in the Humid Tropics.* Washington DC: National Research Council.

Peña, H., Luraschi, M. and Valenzuela, S. 2004. Agua, desarrollo y políticas públicas: La experiencia de Chile. REGA Vol. 1 No. 2 Jul–Dec. pp. 25–50.

Ramos, M.M.G. 1998. *Drenagem Urbana: Aspectos urbanísticos, legais e metodológicos em Belo Horizonte.* Master Dissertation Engineering Faculty, Federal University of Minas Gerais.

Santos, J. 2005. *Water and Health.* Workshops of Integrated Urban Water Management. Iguaçu Falls, April.

Tietenberg, T. 2003. *Environmental and Natural Resource Economics.* Boston: Addison Wesley.

Tucci, C. 2006. *Gestão Integrada de Águas Urbanas.* Workshop de Gestão Estratégica de Recursos Hídricos 4–6 December.

Tucci, C.E.M. 2001. Urban Drainage Management. C. Tucci (ed.) *Humid Tropics Urban Drainage,* Chapter 7. Paris: UNESCO.

Tucci, C.E.M. and Porto, R.L. 2001. Storm hydrology and urban drainage. C. Tucci (ed.) *Humid Tropics Urban Drainage,* Chapter 4. Paris: UNESCO.

Tucci, C.E.M., 2009. Integrated Urban Water Management in Large Cities: A Practical Tool for Assessing Key Water Management Issues in the Large Cities of the Developing World (draft). World Bank. July 2009. 180p.

UN. 2009. Urban and Rural. http://www.un.org/esa/population/publications/wup2007/2007 urban_rural.htm (accessed 01/16/2009).

Weibel, S.R., Anderson, R.J. and Woodward, R.L. 1964. Urban land runoff as a factor in stream pollution. *Journal of the Water Pollution Control Federation*. Washington DC, Vol. 36, No. 7, pp. 914–24.

WHO/UNICEF. 2002. Joint Monitoring Programme for Water and Sanitation WHO and UNICEF. http://www.wssinfo.org/en/337_san_asiaW.html (accessed 10 March 2006).

WHO, 2009. Burden of disease and cost-effectiveness estimates. http://www.who.int/water_sanitation_health/diseases/burden/en/index.html (accessed 23 January 2009).

World Resources Institute. 1992. *World Resources 1992–1993*. Oxford: Oxford University Press.

Wright, A.M. 1997. *Toward a Strategic Sanitation Approach: Improving the Sustainability of Urban Sanitation in Developing Countries*. Geneva: UNDP/World Bank.

Chapter 2

Water supply and wastewater management in the humid tropics

Antônio Domingues Benetti[1] and Luiza Cintra Campos[2]

[1]Hydraulics Research Institute, Federal University of Rio Grande do Sul, Porto Alegre, Brazil
[2]Department of Civil, Environmental and Geomatic Engineering, University College London, United Kingdom

2.1 OVERVIEW

The World Health Organization (WHO) estimates that 24% of the global disease burden and 23% of all deaths can be attributed to environmental risk factors. In children 0–4 years of age, these estimates rise to 36% and 37%, respectively. Many of the risk factors are related to water. For instance, 94% of diarrhoeal disease is attributable to unsafe water, poor sanitation and hygiene (Prüss-Üstin and Corvalán, 2006). The disease burden is highly skewed towards developing countries, most of which are located in the tropics.

Presently, more than 1.1 billion people in the world, most of them in developing countries, are using unsafe water for drinking and hygiene purposes (WHO and UNICEF, 2004). The WHO estimates that it will cost US$37.5 billion over ten years to halve the number of people living without access to safe drinking water (a Millennium Development Goal). The benefit of this action would be 30 million DALYs (Disability-Adjusted Life Years) worldwide (WHO, 2002). One DALY is equivalent to the lifetime loss of one healthy year due to disease. Universal access to safe drinking water (98% coverage), basic sanitation and disinfection at point of use would cost US$486.5 billion, resulting in the gain of an additional 553 million DALYs (WHO, 2002).

This chapter presents aspects of water supply and wastewater management in the humid tropics, including off-site and on-site treatment alternatives. It also includes information on wastewater reuse in urban areas, industrial plants, agriculture and aquaculture.

2.2 WATER SUPPLY INTERACTIONS IN THE HUMID TROPICS

Water supply cannot be taken for granted, even in the wettest regions of the world. In 2005, a severe drought occurred in the Amazon region, one of the wettest areas of the world. Because of the drought, the water level in the Amazon River and its tributaries dropped several metres, resulting in the appearance of the riverbed over a stretch of several hundred metres. The drought also meant that drinking water was scarce and contaminated, causing sickness among the rural population and indigenous Brazilians.

In addition, fish populations, the main source of protein for local people, plummeted, while navigation came to a halt, isolating communities. The dry weather also favoured forest fires (Brasil, 2005; BBC, 2005). These events demonstrate that even the wettest region of the world can fail to provide safe drinking water to its residents.

Alternative options for water supply include communal open wells, tube wells/boreholes, water vendors, public tanker trucks, water kiosks, yard taps, house connections and communal standpipes, yard or roof tanks (Lyonnaise des Eaux, 1998 cited in Stephenson, 2001). But at least half of these are not recommended or not technically viable.

Two-thirds of the world's population, the majority in developing countries, obtain their water from public standpipes, community wells, rivers and lakes, and rainfall collected from roofs (Hinrichsen et al., 1997). In El Salvador, for example, of the percentage without piped water, 8.4% obtain water from public standpipes, 6.9% from tanker trucks, 11.3% from wells, 8.3% from streams or ponds and 7.3% from rainwater or from friends or family (Delgado and Gomez, 2003).

Rainwater harvesting has been practised for more than 4,000 years, due to the temporal and spatial variability of rainfall (UNEP, 2002a). It is used primarily to increase crop production or to provide domestic water supply. Rainfed agriculture is practised on 80% of the world's agricultural land area (UNEP, 2009). Large areas of some countries in Central and South America, such as Honduras, Brazil and Paraguay use rainwater harvesting as a main source of water supply for domestic uses, especially in rural areas (UNEP, 1998). Some other examples in humid tropic countries include Uganda, Singapore, Thailand, Kenya, Indonesia, the Philippines and Bangladesh (UNEP, 2002a).

In the case of communal open wells, a bucket is used to draw water from wells that can easily be contaminated. These wells also dry up during the dry season in the humid tropics, forcing villagers to walk long distances to acquire water. Tube wells or boreholes, where water is drawn from pumps attached to the pipe, have the advantage of abstracting water from deep underground aquifers, but the mechanical pumps needed have to be maintained.

Water vendors who sell water to houses play a central role in water distribution in poor settlements. For example, in Khartoum State, Sudan, water vendors cover 95–98% of water distribution to households where a connection to tap water does not exist (Soman, 2005). Water vendors may use animal-drawn carts or single-axle handcarts as a means of water distribution. Another way of delivering water to homes in developing countries is by tanker truck; these can be public or private sector-owned. In the case of water kiosks, vendors sell water to customers, who have to carry it to their houses by buckets. The city of Onitshan, Nigeria, presents one example of an elaborate water-vending system, which has been created and is operated by the private sector. Approximately 275 tanker trucks purchase water from about twenty private boreholes, which is then sold to households and water kiosks (Whittington et al., 1989).

The condominium water-distribution system has been proposed for small communities as an alternative to the more costly conventional method. The condominium approach enables community involvement in the construction and maintenance of the system. It requires less extension of the pipe network and associated appurtenances by allowing installations within the owners' plots of land. The system has been successfully applied in La Paz, Bolivia and in the city of Parauapebas, located in the Amazon region in Brazil.

Table 2.1 **Water treatment processes and contaminant removal**

Treatment process	Contaminant removal
Pre-treatment	Algal cells, high levels of turbidity, viruses and protozoan cysts.
Sedimentation	Settleable particles.
Dissolved air flotation	Fine low-density particulates, such as flocs containing colour or algae.
Filtration	
Rapid sand filtration	Particles of solid matter, which can include biological contamination and turbidity.
Slow sand filtration	Slow sand filtration is highly effective in removing pathogens (e.g. bacteria, viruses, *Giardia* and *Cryptosporidium* cysts) and dissolved organic matter after pre-oxidation.
Membrane filtration	Nearly all inorganic contaminants (e.g. lead, cadmium, and arsenic), protozoan, bacteria and virus.
Adsorption	Many organic chemicals such as pesticides, solvents, taste- and odour-causing compounds and trihalomethanes. It can also remove some heavy metals and other inorganic compounds.
Chemical oxidation/Disinfection	Pathogens, dissolved organic matter and algae.

In the latter, the cost per connection was estimated at approximately US$8.5, while the estimated cost for the conventional system was US$21.7 per connection (OPS, 2007).

A major concern regarding drinking water in tropical countries is microbial water quality. Several technologies and methods are available to provide effective treatment and disinfection. These are described in the following sections.

2.2.1 Water treatment technologies

Water must be treated using a range of physical, chemical and/or biological processes to remove pollutants harmful to health and prevent water-borne disease outbreaks. Although various pre- and post-treatment technologies exist, filtration forms the heart of the water treatment process. General technologies used for water treatment are summarized in Table 2.1 and described below in more detail. Technologies such as dissolved air flotation, carbon adsorption and membrane filtration have higher costs than simpler processes such as sedimentation, slow sand filtration and disinfection. Since significant parts of developing countries are poor and located within the humid tropics, these simpler technologies are preferred to the more costly alternatives.

Pre-treatment

Pre-treatment involves physical and/or chemical treatment to modify water quality and make it more suitable for treatment by the main unit treatment processes used by the treatment plant.

Physical pre-treatment

The removal of turbidity constitutes the most common form of pre-treatment. Physical pre-treatment processes include roughing filters, micro-strainers, off-stream storage and bank infiltration, each with a particular function and water quality benefit.

Coagulation and flocculation

Chemical coagulation followed by flocculation involves a process in which suspended particles are destabilized and agglomerated to form larger particles that are easier to filter or settle. These are used to increase the effectiveness of other treatment processes and are an integral part of the rapid gravity process. *Moringa oleifera* – a seed from the Moringa tree – acts as a natural coagulant and is therefore appropriate in the humid tropics (Sutherland et al., 1994). However, it may not be appropriate to use chlorine in combination with Moringa oleifera due to the increase in organic matter concentration (CDC and USAID 2009).

Pre-oxidation

Chemical oxidants are applied to water treatment to fulfil a wide variety of objectives including: colour removal, odour and taste control, enhancement of coagulation and flocculation, iron or manganese oxidation and microbial growth. The oxidizing agents most commonly used are ozone, hydrogen peroxide, potassium permanganate, chlorine and chlorine dioxide. There are a number of potential problems with pretreatment oxidation, depending on the raw water quality and the oxidant used:

● It can produce oxidation by-products such as trihalomethanes (THMs), haloacetic acids and bromate.
● Oxidants can lyse algal cells, releasing liver or nerve toxins, or creating objectionable tastes or odours (Chorus and Bartram, 1999).
● Particulate material may interfere with microbial inactivation. Such material protects bacteria and viruses from disinfectants by creating an instantaneous disinfectant demand (preventing the maintenance of a disinfectant residual in subsequent treatment steps) and by shielding the microbe from the oxidant (LeChevallier and Au, 2004).

Sedimentation and dissolved air flotation

Sedimentation

Sedimentation is a physical water treatment process used to settle out suspended particles in water under the influence of gravity. These particles are deposited on the bottom of the settling tank and form a layer of sludge, while the water reaching the tank outlet is in a clarified condition.

Sedimentation in water treatment is generally preceded by chemical coagulation and flocculation. Together, coagulation, flocculation and sedimentation can remove 90–99% of bacteria, viruses and protozoa (LeChevallier and Au, 2004).

Dissolved air flotation

Dissolved air flotation (DAF) is used to remove fine low-density particles such as flocs containing colour or algae from water. It requires energy to produce extremely fine air bubbles. These bubbles attach themselves to the suspended material and thus reduce the density of the particles. This makes the particles more buoyant and they float to the surface.

DAF is particularly effective in removing algal cells and *Cryptosporidium* oocysts (LeChevallier and Au, 2004). In situations where significantly high algal concentrations

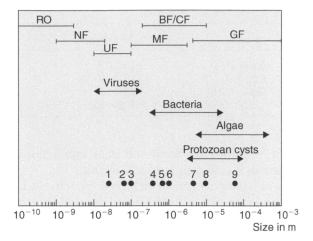

Key: RO: Reverse osmosis; NF: Nanofiltration; UF: Ultrafiltration;
MF: Microfiltration; BF/CF: Bag and cartridge filters; GF:
Granular filtration including slow sand filtration
1: MS2 bacteriophage, 2: Rotavirus, 3: PRDI bacteriophage,
4: *Mycobacterium avium* complex, 5: *Yersinia* spp., 6: Coliform bacteria,
7: *Cryptosporidium* oocysts, 8: *Giardia* cysts, 9: *Balanthidium coli* cysts.

Figure 2.1 **Filter medium pore sizes and the size of microbial particles (with selected microorganisms marked with numbers)**

Source: Stanfield et al., 2003

occur, DAF can be used to remove algae without prior chemical coagulation–flocculation processes (Ives, 2002).

Filtration

Filtration involves the removal of suspended particles by passing water through a porous medium. Figure 2.1 shows the most commonly used filtration processes in drinking-water treatment and how the size of pores in the filter influence the size of microorganism removed.

Rapid sand filtration

Rapid sand filtration (RSF) is the most widely-used process in water treatment today. The process normally uses a combination of chemical (coagulation) and physical (flocculation, sedimentation and filtration) processes to achieve maximum effectiveness and, under optimal conditions, can remove 99.99% or more of the protozoan pathogens. However, without proper chemical coagulation, rapid sand filtration works as a simple strainer and is not an effective barrier against microbial pathogens (LeChevallier and Au, 2004).

Slow sand filtration

Slow sand filtration (SSF) is the oldest water treatment technology. It uses a combination of physico-chemical and biological processes. SSF does not use any chemicals and its by-products are natural compounds that result from biological degradation – carbon dioxide and salts as sulphates, nitrates and phosphates. It is a simple and

Table 2.2 **Membrane pore size, pressure and retention**

Membrane	Pore size (μ)	Pressure (psi)	Substances retained
Microfiltration	0.1–0.2	10–50	Bacteria, protozoa
Ultrafiltration	0.01–0.1	15–200	Colloids, viruses
Nanofiltration	0.001–0.01	70–250	Molecules
Reverse osmosis	<0.001	200–1500	Salts, ions

low-cost process that is ideal for rural areas and small communities, but can also be used for larger communities if sufficient land is available.

The high efficiency of water treatment achieved by slow sand filters is partly explained by the filtration rate (0.1–$0.3\, \mathrm{m\, h^{-1}}$) and the fine effective size of the sand (0.1–$0.3\, \mathrm{mm}$), but is also attributed to biological processes in the layer of slime material (*schmutzdecke*) that accumulates above the sand surface and within the upper layers of the sand bed (Huisman and Wood, 1974). The efficient removal of problematic pathogenic microorganisms, such as *Giardia* cysts and *Cryptosporidium* oocysts (Fogel et al., 1993), by SSF is considered to be a major advantage compared to rapid filtration and other advanced water treatment methods.

Several pre-treatment techniques, such as microstraining, roughing filters and pre-ozonation, have been suggested to overcome the raw water quality limitation of slow sand filters (turbidity <10 NTU and colour <5 CU (Sharpe et al., 1994). SSF combined with pre-treatment (e.g. roughing filters) is known as multi-stage filtration.

Membrane filtration

Membrane filtration is a highly sophisticated process that uses a synthetic polymeric membrane to filter minute particles, including viruses and some ions, out of the solution under pressure. The most commonly-used membrane processes in drinking-water treatment for microbial removal are microfiltration and ultrafiltration. Other membrane processes such as reverse osmosis and nanofiltration, which are used primarily for other purposes, also remove pathogens (see Figure 2.1). Table 2.2 summarizes the pore size, pressure and substances that are retained in the membranes.

In practice, membrane filtration application is still limited in the tropics because of capital, operation and maintenance costs. However, it has been applied in Brazil's semi-arid region to treat brackish water for community water supplies. The application of membrane processes is expected to increase over the next few years due to the increased pollution of drinking-water sources and reduction in the cost of membranes (Global Markets Direct, 2009).

Adsorption

Adsorption is a process whereby soluble molecules (adsorbate) are removed by attachment to the surface of a solid substrate (absorbent). Van der Waals forces are the predominant mechanism, but chemical or electrical attraction is also important.

The most widely used adsorbent is activated carbon, which is either added to water as a powder or slurry, or as a granule that is housed in a simple filter or column through which contaminated water is passed. In drinking-water treatment, activated carbon is used to remove colour, taste and odour-causing compounds, as well as other organic residuals, such as cyanotoxins, pesticides and chlorination disinfection by-products (e.g. trihalomethanes (THMs)).

Disinfection

Water disinfection can be accomplished by chemical and physical processes. The most common chemical disinfectants are chlorine, chlorine dioxide and ozone. Other chemical disinfectants include hydrogen peroxide, iodine, bromine and silver. Physical disinfectants may include SSF, membranes, artificial UV radiation and solar radiation. These can be used either singularly or increasingly in combination to maximize the efficiency of microbial inactivation.

Chlorine

Chlorine is the oldest and most widely-used disinfectant. It has several characteristics that support its application: it is effective against a wide spectrum of pathogenic organisms; it has residual action in the distribution system; it requires only simple dosage equipment; it is available even in remote places in developing countries; and it is an efficient and relatively low-cost option (Solsona and Méndez, 2002).

Chlorine can be applied as a gas (Cl_2), liquid (sodium hypochlorite, $NaClO$) or solid (calcium hypochlorite, $Ca(ClO)_2$). These compounds dissociate in water to form hypochlorous acid ($HOCl$) and hypochlorite ion (OCl^-) depending on the pH. $HOCl$ has a bactericidal power about 80% greater than (OCl^-). These chemicals inactivate microorganisms by interfering with the enzymatic activities of bacteria and viruses.

When chlorine is added to water, it reacts with dissolved organic compounds to form a variety of chloro-organic compounds. Of particular concern in the context of drinking-water disinfection by chlorination are the THMs, which are formed from the reaction between free chlorine and naturally occurring humic substances in water (Casey, 1997). Some THM compounds (e.g. chloroform) are suspected carcinogens. The total THM concentration in chlorinated drinking water is a function of the organic precursor concentration and the chlorine dose used. When chlorine is added to water containing natural or added ammonia, the ammonia reacts with hypochlorous acid to form various chloramines. Chloramines are effective disinfectant agents but are not as potent as hypochlorous acid (Casey, 1997).

In remote areas where a continuous supply of chlorine cannot be guaranteed, sodium hypochlorite can be produced on site using electrolysis. A 3% sodium chloride solution (30 g NaCl/litre) submitted to electrolysis will generate 400 litres per day of solution containing 5 to 7 g of active chlorine, sufficient to provide safe water to a community of 5,000 inhabitants (Solsona and Méndez, 2002).

Chlorine dioxide

Chlorine dioxide (ClO_2) has higher disinfection power than other chlorine compounds. It can improve the organoleptic quality of the drinking water by destroying substances that cause taste and odour in water. It also reacts efficiently with disinfectant by-product precursors. Chlorine dioxide has to be produced on site. Because of the complexity of its production and relatively high cost, the application of ClO_2 has

Table 2.3 Ranking of disinfectants based on bactericidal efficiency, stability and dependence on pH

Disinfectant	Bactericidal efficiency	Stability	pH dependence
Ozone	1°	4°	Small
Chlorine dioxide	2°	3°	Small
Chlorine	3°	2°	Medium
Chloramines	4°	1°	Small

Source: Solsona and Méndez, 2002

been limited to medium to large systems. Chlorine dioxide does not dissociate, it destroys bacteria by affecting their cell membranes. In addition to bacteria, ClO_2 can also effectively destroy enterovirus and *Cryptosporidium*. The ions chlorite (ClO_2^-) and chlorate (ClO_3^-) are the best-known disinfection by-products of chlorine dioxide.

Ozone

Ozone (O_3) is a powerful oxidant, capable of eliminating pathogenic microorganisms and compounds that give colour and an unpleasant odour to water. Because of its insta-bility, ozone must be produced on site and used almost immediately. Unlike chlorine, ozone does not leave residues when it is used for oxidation and disinfection in the water distribution system. Disinfection with ozone occurs by reaction with protoplasmic material. As with chlorine, ozone can react with compounds present in water to form disinfection by-products. Ozone does not form halogenated by-products (e.g. THMs) during the disinfection process when natural organic matter (NOM) is present, but it does form a variety of organic and inorganic compounds. However, if bromide ions are present in the raw water, halogenated by-products may be formed, which appear to present a greater health risk than non-brominated by-products (US-EPA, 1999). Table 2.3 compares bactericidal efficiency, stability and the effect of pH on the action of chlorine, ozone, chlorine dioxide and chloramines. It can be seen that the higher the bactericidal capacity of the disinfectant, the lower its stability.

Ultraviolet radiation

Ultraviolet (UV) is produced by mercury-vapour light bulbs similar to fluorescent lights. UV radiation has a bactericidal effect. It has received great attention in higher-income countries since the findings on the production of disinfection by-products with chlorine use. UV can be used for drinking-water and wastewater disinfection. It is a physical method where short wavelengths, similar to those of sunlight, destroy genetic material from bacteria and viruses without chemical or physical changes in water. Important vari-ables with respect to UV disinfection are: wavelength, water quality, radiation intensity, type of microorganisms and contact time.

Other disinfection methods

Other chemicals that inactivate microorganisms are bromine and silver. Bromine is a halogen that acts similarly to chlorine, whilst silver is effective against bacteria, but not against viruses.

Sodium Dichloroisocyanurate (NaDCC) is a compound that releases high concentrations of active chlorine (60%) as hypochlorous acid (HOCl). Unlike other chlorine compounds, it does not cause taste and odour problems. It is very stable and can be stored for more than five years – much longer than other disinfectants. Because of its advantages, use of this compound is expected to increase in the future.

Advanced oxidation processes generate hydroxyl radicals (OH), which are powerful oxidants. Hydroxyl radicals are used to oxidize chemicals that are recalcitrant or difficult to destroy (WHO, 2004). Hydroxyl radicals can be produced by combining UV with hydrogen peroxide (H_2O_2/UV) or ozone (O_3/UV), ozone with hydrogen peroxide (O_3/H_2O_2), and UV (O_3/H_2O_2/UV). Hydroxyl radicals can also be produced by photocatalysis, an advanced oxidation process in which hydroxyl production is catalyzed by titanium oxide (TiO_2).

Table 2.4 presents a summary of the disinfectants and their main characteristics. From Table 2.4, chlorine, solar water disinfection (SODIS) and SSF are the most acceptable processes for developing or low-income countries due to their low capital, operation and maintenance costs, as well as the lower skill requirements for the operator. Although SSF is a removal process, it can be considered as a disinfectant method due to its efficiency of pathogen removal. SODIS is discussed below in Section 2.2.2.

2.2.2 Household water treatment processes

Because of potential contamination in the distribution system, it may be necessary to treat water again at the point-of-use in order to protect health. There are many types of household water treatment technologies available, including membrane filters, activated carbon filters, slow sand filters, ceramic filters, boiling and other physical and chemical methods.

The filters essentially employ the same technological approaches as large-scale treatment systems, but on a much smaller scale. The effectiveness for particle and microbial removal varies widely depending on the type of microbe as well as the type and quality of the filtration media. Therefore, testing to ensure that the water is of a satisfactory quality is a useful safety precaution. It is important to note that at some point, all devices require maintenance, but some more than others.

However, none of these filters can address every water quality problem. In general, household technologies are used to control turbidity and to reduce levels of organic contaminants, arsenic, nitrate, microorganisms (including cysts) and many other contaminants (Lahlou, 2003). Aesthetic aspects, such as taste, odour or colour, can also be improved with household treatment.

In situations where the water is contaminated by pathogens, three treatment methods are available:

1. The simplest and oldest method is to boil the water. This method can kill most types of disease-causing organisms. Usually, the water is boiled for one minute and can be consumed after cooling (US-EPA, 2006). One disadvantage of boiling is the consumption of energy, estimated as 1 kg of wood per litre of water (Sobsey, 2002). It requires great care because of the risk of accidents, particularly burns to children.
2. Water can be disinfected using household chlorine (bleach) and iodine. Both will destroy some, but not all, types of disease-causing organisms. Deciding on the right

Table 2.4 Characteristics of disinfectants

		Chlorine gas	Ozone	Chlorine dioxide	Ultraviolet	Membranes	Slow sand filtration	Solar disinfection
Removal efficiency	Bacteria	H	H	H	H	H	H	H
	Viruses	H	H	H	H	H	I	H
	Protozoa	L	I	I	L	H	H	L
	Helminths	L	H	L	L	H	H	L
Water quality influence	pH	H	L	L	L	H	L	L
	Turbidity	I	H	L	H	H	I	H
	Organic matter	L	L	L	L	I	L	I
Disinfection by-products		Yes	L	L	No	No	No	No
Residual in distribution system		Yes	No	No	No	No	No	No
Taste and odour production		Yes	No	No	No	No	No	No
Flow rate range		L-I-H	L-I-H	I-H	L-I-H	I-H	L-I	L
Capital costs (equipments)		L	I-H	I-H	L-I	H	I	L
Operation & maintenance costs		L	H	H	L-I	H	L	L
Operator skills		L-I	H	H	L-I	H	L	L
Chemicals requirement		Yes	No	Yes	No	No	No	No
Energy requirement		L	H	H	I	H	No	No

Legend: L – Low; I – Intermediate; H – High
Source: Solsona and Méndez, 2002

amount can be difficult, since it depends on the substances in the water which react with the disinfectant, and which may vary from season to season (Skinner and Shaw, date unknown).

3. SODIS is a simple and inexpensive method for the disinfection of microbiologically contaminated water. SODIS uses solar energy to deactivate pathogens that are vulnerable to the effects of solar radiation in the spectrum of UV-A light (wavelength 320–400 nm) combined with increased water temperature (Meierhofer and Wegelin, 2002). SODIS is therefore highly appropriate for small communities located in warm regions.

A variety of equipment can be used for solar disinfection – solar panels, solar cooking stoves, solar concentrators and solar distillers. A simple technique for water disinfection is exposing a plastic soft drink bottle, half-filled with water, to sunlight for several hours. The treatment efficiency can be improved if the plastic bottles are exposed on sunlight reflecting surfaces such as aluminium or corrugated iron sheets. SODIS has been shown to remove faecal coliforms and *Vibrio cholerae*, and is currently used in Asian, African and South American countries (Sommer et al., 1997). Although there have been allegations of migration of carcinogenic substances from reused PET bottles in drinking, a recent study (Schimid et al. 2008) shows that there is no reason for concern.

2.3 WASTEWATER COLLECTION, TREATMENT AND REUSE

2.3.1 Off-site sanitation systems

Wastewater in urban areas can be collected by pipes or can be disposed of on-site. If it is collected by pipes, it can be separated or combined with run-off. Where conveyance systems are available, commercial, institutional and industrial wastewater may be discharged to them. Figure 2.2 shows the flow contributions to urban conveyance systems.

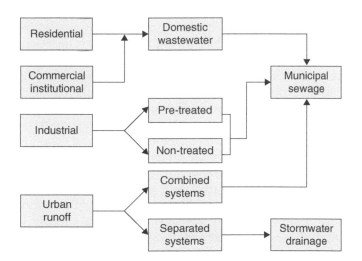

Figure 2.2 **Wastewater and stormwater in urban conveyance systems**

Source: adapted from Veenstra et al., 1997

Conventional sewerage is costly and its sustainability has been questioned (Matsui et al., 2001). Low-cost sewerage technologies such as condominium arrangements are an alternative solution. In this system, pipelines are placed within the dweller's plots of land. The pipes can be settled at lower depths where they are not exposed to loads from vehicles. Wastewater collected from several houses is transported to a single access chamber, from which it is connected to the public sewer. Thus, only one connection is required for several houses. This feature reduces costs because a single line of sewer can receive sewage contributions from an entire block. Although condominium arrangements have been used in various countries in Latin American such as Brazil, Bolivia and Peru (Lampoglia and Mendonça, 2006; WSP, 2007), conventional sewerage is still in use worldwide.

In many cities, wastewater and stormwater are collected in a single pipeline. During storms, the capacity of the pipes and treatment plants is exceeded, and an overflow is produced. The Combined Sewer Overflow (CSO) may severely affect the quality of the receiving body of water because of the loads of oxygen-demanding substances (organic matter and ammonia); nutrients (nitrogen and phosphorus); toxicants (ammonia, metals and organic micropollutants); health-impairing pathogens (faecal bacteria); and physical contaminants (temperature, suspended solids and chloride).

The shock loads can also have serious impacts on the performance of wastewater treatment plants, for instance, the sludge blanket washout from Upflow Anaerobic Sludge Blanket (UASB) reactors (Salazar and Benetti, 2007) and final clarifiers (Lijklema et al., 1993). Environmental effects on the receiving water body include dissolved oxygen reduction, biomass accumulation, enrichment, toxicity, temperature rise, blanket formation by solids deposition and contamination with pathogens. Beneficial uses affected by contamination are water supply, bathing, recreation, fishing, industrial water supply, irrigation and ecosystem preservation.

Treatment alternatives

Treatment processes are classified according to the method that acts to promote the removal of the contaminant. In physical unit operations, physical forces are applied to remove pollutants. In chemical and biological unit processes, the removal of contaminants is achieved by chemical or biological reactions. Unit operations and processes are combined to provide different levels of treatment. These levels are classified as preliminary, primary, secondary and tertiary, with each level providing additional contaminant removal. Some additional classification includes advanced primary, secondary with nutrient removal and advanced. Table 2.5 describes the types of contaminants that are removed during different levels of treatment.

Traditional wastewater treatment technologies are sometimes beyond the investing capacity of developing countries. However, there are treatment alternatives that are capable of achieving high levels of treatment with relatively low capital and operational costs. One example is Waste Stabilization Ponds (WSP) comprising anaerobic, facultative and maturation ponds in series. This system is effective for Biochemical Oxygen Demand (BOD) and pathogen removals. Pond effluents can meet the WHO quality standards for unrestricted irrigation. This makes this treatment technology highly recommended for regions where reuse is being considered. Stabilization ponds require careful operation, particularly with respect to odour production and mosquito breeding, which can undermine public acceptance of the facility.

Table 2.5 Contaminant removal at levels of wastewater treatment

Treatment level	Contaminant removal	Examples of unit operations or processes
Preliminary	Rags, sticks, plastic, papers, grit	Screening, grit removal
Primary	Settleable suspended solids and organic matter	Sedimentation
Advanced primary	Enhanced removal of suspended solids and organic matter	Chemical coagulation–sedimentation
Secondary	Colloidal and dissolved biodegradable organic matter	Stabilization ponds, aerated lagoons, activated sludge, trickling filters, rotating biological contactors
Secondary with nutrient removal	Nitrogen and/or phosphorus	Nitrification–denitrification, overland flow, chemical precipitation, biological phosphorus removal
Tertiary	Residual suspended solids, pathogens	Chemical precipitation, filtration, chlorine disinfection, maturation ponds
Advanced	Dissolved solids and organics	Carbon adsorption, membrane filtration, advanced oxidation

Source: Crites and Tchobanoglous, 1998; Aoki et al., 2004

WSP performance is often defective due to problems such as under-design, short-circuiting, incorrect loading, poor operation and maintenance, and adverse environmental factors, leading to poor quality effluents (Lloyd et al., 2006). With the exception of maturation ponds, effluents from stabilization lagoons have a high concentration of suspended solids, composed mostly of algae. This may constitute a problem where the receiving body of water is prone to eutrophication, such as lakes and estuaries. The enrichment of water bodies with nutrients causes algae and cyanobacteria blooms with the potential production of compounds that are toxic to humans and animals (Chorus and Bartram, 1999).

Another treatment technology largely used in warm climates is the UASB reactor (van Haandel and Lettinga, 1994; Aiyuk et al., 2006). This reactor can replace primary sedimentation tanks, sludge thickening and digestion units. Depending on the effluent quality standards, the UASB reactor must be followed by another treatment technology, such as a trickling filter, or facultative and maturation ponds.

Wetlands, defined as lands in which the water table is at or above the soil surface during part of the year, have been used to treat wastewater from urban, industrial and agricultural sources. These are very effective in removing BOD, suspended solids, nitrogen and pathogens, as well as significant levels of metals and trace organics from wastes (Reed et al., 1995). Constructed wetlands are a promising wastewater treatment technology for developing countries, being particularly effective in tropical climates. They require only a low level of technology investment and infrastructure (Coughanowr, 1998). In Calcutta, India, wetlands are integrated into a resource recovery system in which effluent is used for aquaculture and agriculture, yielding 7,000 to 8,000 tonnes of fish per year. Another example is the wetland constructed to remove nutrients from agricultural run-off before entering the Everglades National Park, in Florida, US. Constructed wetlands for wastewater treatment have also been used successfully in El Salvador, Nicaragua and Honduras (CAD et al., 2007).

2.3.2 On-site sanitation systems

In 2000, about 2.4 billion people lacked access to basic sanitation; the majority of them were living in Africa and Asia. According to the report, 'The Global Water Supply and Sanitation Assessment 2000', about 2 billion people live in rural areas (WHO and UNICEF, 2000). The report also lists sanitation technologies accepted as adequate as: connection to a public sewer, connection to a septic system, simple pit latrine, ventilated improved pit latrine and pour-flush latrine. The first technology, connection to a public sewer, requires that household wastewater be linked to a sewerage system, and is classified as an 'off-site' system. Bucket, public and open latrines are not considered proper waste disposal methods.

Septic systems are used in houses that have water connections and full plumbing. Tanks are placed in the yards of the houses, and are built below the soil level. Wastewater from the household is transported by pipes to the septic tank, where it remains for one to three days. The tank effluent is moved for disposal in a soil absorption system such as leach pit or trench (Feachem et al., 1983).

A pit latrine is an on-site dry system in which a hole is dug in the ground, and may be lined or unlined. A slab is placed over the hole that receives the excreta. When properly located with respect to the well and groundwater level, the pit latrine can be a low-cost alternative for excreta disposal. In the ventilated improved pit latrine, a vent pipe is connected to the pit; this almost entirely eliminates the disagreeable odour released from the pit.

In the pour-flush toilet, a water seal prevents contact with odour and insects from the pit. Excreta are flushed from the toilet pan into a pit by adding 2 to 3 litres of water as an excreta carrier. The pit is built in the same way as the pit latrine, and care must be taken to prevent groundwater contamination.

Another on-site method that can provide safe excreta disposal is the composting toilet, a type of ecological sanitation (see Section 2.3). In one type, two adjoining chambers are built above ground level. Aerobic decomposition of the faecal sludge is achieved by allowing air to pass through the sludge. The air is exhausted from the chamber through a vent located in the opposite wall. The two composting chambers are used alternately. While one is left for composting, the other is receiving excreta. This method provides mature compost that can be used in the garden (UNEP, 2002b). A similar type of system, called the Sulabh Sauchalayas has been extensively built in India as part of the Sulabh Sanitation Movement. In the city of Shirdi, India, the movement built what is acknowledged as the world's largest public toilet and bathing facility (SISSO, 2007).

2.3.3 Wastewater reuse

Within twenty-five years, it is estimated that two-thirds of the world's population will be living in water-stressed countries mainly located in West Asia, North Africa or sub-Saharan Africa (UNEP, 2002c). This forecast accentuates the need for wise water management, including the reuse of treated wastewater. Although countries in the humid tropics might have very high water availability ($>20,000\,m^3$ per capita per year), wastewater reuse can be an attractive alternative to more conventional treatment technologies because it contains substances, such as nitrogen, phosphorus and organic carbon, that are useful for certain applications. For instance, the use of treated wastewater in agriculture, landscaping and gardening can reduce the use of industrial fertilizers, while maintaining soil fertility. It can also save costs due to reduced water consumption and the sizing of the treatment facilities. Besides agriculture, water reuse has applications in industrial plants, urban buildings, recreation, aquaculture and groundwater recharge (Asano et al., 2007).

In order to design sustainable wastewater reuse applications, it is necessary to evaluate the potential health risks associated with the reuse and the measures needed to minimize them. Additionally, the water quality of the treated wastewater must meet the requirements for the intended use (Aoki et al., 2004).

Wastewater reuse in agriculture

Agriculture is the largest user of water, using about 70% of the freshwater drawn from rivers, lakes and groundwater (UNEP, 2002b). Agricultural irrigation is implicated in the disappearance of ecosystems and the shrinking of water bodies such as the Aral Sea in Northwest Asia and Lake Chad in West Africa (Aoki et al., 2004).

The application of human excreta to rice fields was common in ancient China (Landes, 1999). This practice improved agricultural yields and food production but, at the same time, created an occupational hazard for farmers due to helminth and schistosome infestation.

The main advantages of wastewater reuse in agriculture are the reduction in industrial fertilizer use, the maintenance of soil fertility, and the prevention of erosion. In order to minimize the risks associated with the presence of pathogens in wastewater, WHO published guidelines for the safe use of wastewater in agriculture (WHO, 2006a).

WHO has established 10^{-6} DALY per person per year as the recommended health protection level for food crops irrigated with treated wastewater. For unrestricted irrigation, this level of protection requires a pathogen reduction of 6 and 7 log units for the consumption of leaf and root crops, (i.e. lettuce and onions), respectively. This level of protection is to be achieved for rotavirus, which also ensures sufficient protection against bacterial and protozoan infections.

The 6–7 log units pathogen reduction can be reached by using a combination of health protection measures. For instance, wastewater treatment can achieve a 1–6 log unit reduction, depending on the process; product washing with water, another 1 log unit; peeling the vegetables, another 2 log units, and so on. Table 2.6 shows options for the reduction of pathogens to achieve the health protection of $\leqslant 10^{-6}$ DALY per person per year.

In Table 2.6, option A, the 6–7 log unit reduction, is obtained by wastewater treatment (4 log units), pathogen dieoff between the last irrigation and vegetable consumption (2 log unit reduction), and produce washing (1 log unit). Alternatives B through E demonstrate other options to reach the 6–7 log unit reduction in unrestricted irrigation.

With respect to helminth infections, the microbial reduction target was based on epidemiological investigations. Field studies conducted in Brazil have demonstrated that helminth eggs are not detected in crops irrigated with a facultative pond effluent containing less than 0.5 eggs per litre (WHO, 2006a). For this reason, a target of $\leqslant 1$ egg per litre of treated wastewater is recommended for unrestricted irrigation. However, additional measures might be necessary to protect children under 15 eating uncooked food crops brought home directly from the fields.

For labour-intensive restricted crop irrigation, in order to achieve the health-based target of $\leqslant 10^{-6}$ DALY per person per year, a 4 log unit pathogen reduction is required

Table 2.6 **Combinations of health protection measures for pathogen reduction to achieve health protection of $\leqslant 10^{-6}$ DALY per person per year**

Health protection measure for pathogen reduction	Unrestricted irrigation					Restricted irrigation		
	A	B	C	D	E	F	G	H
Treatment	4	3	2	4	7	4	3	1
Die-off	2	2	0	0	0	0	0	0
Washing of produce	1	1	4	2	0	0	0	0
Drip irrigation, high crops	0	0	0	0	0	0	0	0
Drip irrigation, low crops	0	0	0	0	0	0	0	6
Subsurface irrigation	0	0	0	0	0	0	0	0
Total $\log_{10}$ pathogen reduction	7	6	6	6	7	4	3	7

Source: WHO, 2006a

(alternative F). For highly-mechanized agriculture, the pathogen reduction is 3 log units (alternative G). Option H in Table 2.6 represents the alternative where wastewater is treated in a septic tank with subsurface irrigation through the soil absorption system. Pathogen reduction in the septic tank is 0.5 log units, while the remaining 6.5 log units are achieved by the soil.

Table 2.7 presents wastewater treatment alternatives to achieve the specified log unit reduction of microorganisms.

Wastewater reuse in aquaculture

Aquaculture is the practice of growing fish, algae and macrophyte in ponds. Wastewater and excreta are sources of nutrients and water, and were used for centuries in China, ancient Egypt and in Europe in the Middle Ages, to fertilize crops and fishponds (Feachem et al., 1983).

In a fishpond, nutrients in wastewater promote phytoplankton and zooplankton growth, resulting in an active food chain where fish like carp and tilapia can thrive. However, there are health issues related to this practice that must be addressed in order to prevent diseases. Three important aspects are:

● Fish can carry human pathogens in their intestines and on their body surfaces, thereby posing risks to people who handle, prepare or eat the fish.
● Fish can be intermediate hosts for helminths.
● Other organisms in the pond, in addition to the fish, can host helminths.

WHO has recently published guidelines for the safe use of wastewater and excreta in aquaculture (WHO, 2006b). The guidelines present health-based targets that promote levels of health protection to an exposed population. The level of protection can be achieved by a combination of management practices and microbiological water quality. Practices protect consumers, workers and the community. Among the recommended protection measures are wastewater treatment, disease vector control, produce restriction, cooking and personal protective equipment. The recommended microbial water quality is less than 10^4 *E. coli* per 100 mL, which can be achieved through waste stabilization ponds, sequential batch-fed wastewater storage and

Table 2.7 Log unit reduction or inactivation of pathogens achieved by wastewater treatment technologies

Treatment processes	Log unit pathogens removal			
	Virus	Bacteria	Protozoan (oo)cysts	Helminth eggs
Low-rate biological processes				
Waste stabilization ponds	1–4	1–6	1–4	1–3
Wastewater storage and treatment reservoirs	1–4	1–6	1–4	1–3
Constructed wetlands	1–2	0.5–3	0.5–2	1–3
High-rate processes				
Primary treatment				
Primary sedimentation	0–1	0–1	0–1	0–<1
Chemically enhanced primary sedimentation	1–2	1–2	1–2	1–3
Anaerobic upflow sludge blanket reactor	0–1	0.5–1.5	0–1	0.5–1
Secondary treatment				
Activated sludge + secondary sedimentation	0–2	1–2	0–1	1–<2
Trickling filter + secondary sedimentation	0–2	1–2	0–1	1–2
Aerated lagoon + settling pond	1–2	1–2	0–1	1–3
Tertiary treatment				
Coagulation/flocculation	1–3	0–1	1–3	2
High-rate granular or slow-rate sand filtration	1–3	0–3	0–3	1–3
Dual-media filtration	1–3	0–1	1–3	2–3
Membranes	2.5–>6	3.5–>6	>6	>3
Disinfection				
Chlorination (free chlorine)	1–3	2–6	0–1.5	0–<1
Ozonation	3–6	2–6	1–2	0–2
Ultraviolet radiation	1–>3	2–>4	>3	0

Source: WHO, 2006a

treatment reservoirs, activated sludge followed by filtration, and disinfection or pol-
ishing ponds (WHO, 2006b).

In Lima, Peru, a fifteen-year project supported by the World Bank, the United
Nations Development Programme (UNDP), Deutsche Gesellshaft für Technische
Zusammenarbeit (GTZ), the Pan American Health Organization (PAHO) and the
Peruvian government evaluated fish farming using effluent from stabilization ponds to
grow tilapia. Results from this study showed that the ponds provided safe water qual-
ity for aquaculture (Moscoso, 1998).

Wastewater for non-potable reuse

Some urban water applications do not need water of drinking-water quality. The use
of treated domestic wastewater can conserve limited water resources, particularly in
large cities. Secondary treated wastewater followed by sand filtration and disinfection
has been used in Japan and other countries for non-potable uses, such as toilet flushing, car washing, park lawn irrigation and fire-fighting, particularly in concentrated
areas of large cities (Aoki et al., 2004). As illustrated in Figure 2.3, a dual distribution
system was implemented to supply treated wastewater for use in toilet flushing. The

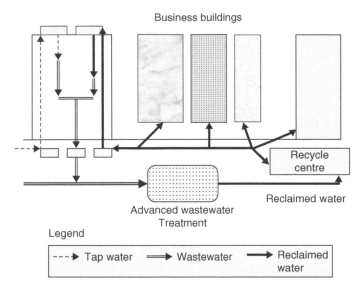

Figure 2.3 **Wastewater recycling scheme in Shinjuku area in Tokyo, Japan**

Source: adapted from Aoki et al., 2004

system has operated since 1984, supplying up to 8,000 m^3 day^{-1} to twenty-five business buildings in Tokyo.

However, the management of a dual distribution system with potable and non-potable water requires great care. Cross-connection between systems and incorrect use of the non-potable system as drinking water can expose the population to disease-causing microorganisms. Other problems related to reuse of treated wastewater are corrosion and biofilm build-up in the pipes and reservoirs, since reclaimed water contains more salt and organic matter than drinking water.

Wastewater for nutrient recycling

Wastewater is composed of urine, faeces and greywater. The nutrient content of urine comprises about 66% and 50% of the nitrogen and phosphorus loads of sewage, respectively (Matsui et al., 2001). Many wastewater treatment plants were not designed to remove nutrients from sewage. Effluents from these plants contribute to the growing problem of eutrophication of lakes, reservoirs, bays, estuaries and enclosed seas. Eutrophication of water bodies promotes algal and cyanobacterial blooms, which may produce toxins that are dangerous to humans and aquatic life (Chorus and Bartram, 1999).

Matsui et al. (2001) suggest that the separation of urine from faeces in the conventional sewage collection system can help to reduce eutrophication and save resources through the recycling of nutrients present in urine. In urine separation toilets, urine is collected separately from faeces and it is conducted to a storage tank near the building. From this tank, urine can be collected and used as fertilizer after dilution. For developing countries that lack the financial resources to provide basic sanitation for urban residents, an 'Ecological Sanitation' approach is available. In this system, a dry toilet diverts urine and faeces from greywater at source. The resulting greywater can be treated and discharged nearby, without the need for an extensive pipeline network. Urine can be collected and transported to the processing plant while faeces are treated on site for pathogen removal. This approach treats excreta as a resource as opposed to a waste (Matsui et al., 2001).

2.4 INDUSTRIAL WASTEWATER COLLECTION, TREATMENT AND REUSE

Many developing countries in the tropics have industrial plants with pollutant emissions similar to those from industrial plants in developed countries. For instance, chemical, oil refining, petrochemical, steel, pulp and paper, automobile and textile plants, among others, are located in several regions of Brazil. The mining of coal, gold, iron and other metals are also important economic activities in the country (Benetti et al., 2004).

In addition, Brazil is a great producer of grains, which are cultivated under modern intensive agriculture practices, characterized by the use of large amounts of fertilizers, pesticides and water. In this way, developing countries in the tropics face a compound pollution problem since they lack the financial resources to collect and treat municipal wastewater and, at the same time, have to deal with industrial and agriculture pollution.

Large industrial plants usually treat their own effluents. Small and medium-sized industries discharge in public collection systems, if they are available, after some pretreatment. In some areas, industrial plants organize a consortium and build a central wastewater treatment plant to receive effluents from several plants.

Many industrial plants are now reusing their treated effluent. Among the reasons described by executives for adoption of this policy are: saving costs, saving water that is being increasingly regulated, and improving the industry's public image.

Water bought from industries can be expensive, especially if they are heavy water users. They may try to minimize costs by using reclaimed water, which might come from treated municipal wastewater or water recycling within the plant. Water is reused for purposes such as processing, washing and cooling. Steel, petroleum refining, chemical, and pulp and paper are major users of reused wastewater (Metcalf and Eddy, 2003).

Cooling tower make-up water is the major water reuse application in many industries. In cooling towers, warm water is recirculated and cooled by heat exchange. Treated wastewater is used to replace water that has evaporated or has been discharged from the cooling system. Wastewater intended for use in cooling towers must be treated to prevent corrosion, scaling, fouling and biological growth on heat-exchanger surfaces. Other applications of industrial water reuse are cleaning, and lawn and garden irrigation.

Opportunities for water recycling in industrial plants can be identified using the Cleaner Production (CP) strategy. This approach aims to prevent pollution; reduce water, energy and material resources; and minimize waste generation in the production process, thereby reducing costs and environmental impacts (UNEP, 2004). The sustainable consumption and production concept considers the entire life cycle of products in order to reduce their environmental impact. It considers the following stages of the product life cycle: product design, selection of raw materials, production process, consumer use and recycling of discarded products (UNEP, 2004).

The UNIDO/UNEP National Cleaner Production Centre Programme established twenty-four centres worldwide. The purpose of the programme is to build local capacity for the implementation of cleaner production techniques in developing countries and economies in transition.

2.5 CONCLUDING REMARKS

Most people without access to safe drinking water and basic sanitation live in developing countries, many of which are located in the humid tropics. This region is characterized by warm weather and abundant rainfall, sometimes associated with extreme events such as cyclones and monsoons.

While the availability of safe drinking water and the adequate disposal of wastewater are universal needs, the technology associated with the attainment of those goals may vary among regions. For instance, wastewater treatment based on soil disposal such as slow or rapid-rate infiltration are probably inadequate for the humid tropics because the land is soaked with water during part of the year. However, aquatic systems such as wetlands and stabilization ponds are more suitable for those conditions.

This chapter reviewed technologies that can be used to produce potable drinking water and remove contaminants from wastewater. A range of technologies is available and the choice of the proper processes depends on factors such as the density of the population to be served (urban vs. rural areas), the raw water quality, and the availability of water distribution and sewerage systems. Disinfection continues to play a special role in preventing the transmission of water-borne infections, which is a major public health concern and problem in developing countries. Chlorine and solar disinfection are two technologies that are applicable to the humid tropics.

The section on reuse considered the increasing emphasis on resource recovery and sustainability. Wastewater contains valuable nutrients that can be used in agriculture and aquaculture, while non-potable water may suit some urban and industrial uses.

Some technologies that were initially developed for regions with different weather and climate patterns are also used in the humid tropics. Designers often use data and values from the literature written for temperate areas of the world, since local data are not yet available. For this reason, research and technology development focused on the humid tropics needs to be supported, to allow for better use of the particular features of the region.

REFERENCES

Aiyuk, S., Forrez, I., Lieven, K., Haandel, A. and Verstraete, W. 2006. Anaerobic and complementary treatment of domestic sewage in regions with hot climates – a review. *Bioresource Technology*, Vol. 97, No. 17, pp. 2225–41.

Aoki, C., Memon, M.A. and Mabuchi, H. 2004. *Water and Wastewater Reuse. An Environmentally Sound Approach for Sustainable Urban Water Management*. Osaka/Shiga: United Nations Environment Programme (UNEP).

Asano, T., Burton, F.L., Leverenz, H.L., Tsuchihashi, R. and Tchobonoglous, G. 2007. *Water Reuse. Issues, Technologies, and Applications*. New York: McGraw-Hill.

Benetti, A.D., Lanna, A.E. and Cobalchini, M.S. 2004. Current practices for establishing environmental flows in Brazil. *River Research and Applications*, Vol. 20, pp. 427–44.

BBC News. 2005. Amazon drought emergency widens. 15 October 2005. http://news.bbc.co.uk/2/hi/américas/4344310.stm.

Brasil, K. 2005. Na seca, crianças morrem com diarréia. *Jornal Folha de São Paulo*, 22 October. 2005, p. C5.

Casey, T.J. 1997. *Unit Treatment Processes in Water and Wastewater Engineering*. UK: John Wiley & Sons.

CDC and USAID (2009). Preventing Diarrheal Disease in Developing Countires: Simple options to remove turbidity. Factsheet January 2009, http://www.ehproject.org/PDF/ehkm/turbidity 2009.pdf (accessed on 4th February 2010).

Chorus, I. and Bartram, J. (eds) 1999. *Toxic Cyanobacteria in Water: A Guide to their Public Health Consequences, Monitoring and Management*. London: Taylor & Francis. (Published on behalf of the World Health Organization.)

Cooperación Austriaca para el Desarrollo (CAD), Agencia Suiza para el Desarrollo y la Cooperación and Water and Sanitation Program. 2007. *Biofiltro: Una Opción Sostenible para el Tratamiento de Aguas Residuales en Pequeñas Localidades*. Programa de Agua y Saneamiento Región América Latina y el Caribe, Oficina Sub-Regional para America Central. Tegucigalpa, Honduras.

Coughanowr, C. 1998. *Wetlands of the Humid Tropics*. Paris: UNESCO, International Hydrological Programme.

Crites, R. and Tchobanoglous, G. 1998. *Small and Decentralized Wastewater Management Systems*. New York: WCB/McGraw-Hill.

Delgado, R. and Gomez, A.L. 2003. *El Salvador: Soyapango Women Fight for Public, Quality Drinking Water*. http://www.citizen.org/cmep/Water/gender/articles.cfm?ID=10793 (accessed 21 February 2008).

Feachem, R.G., Bradley, D.J., Garelick, H. and Mara D.D. 1983. *Sanitation and Disease. Health Aspects of Excreta and Wastewater Management*. Washington DC: World Bank.

Fogel, D., Isaac-Renton, J., Guasparini, R., Moorehead, W. and Ongerth, J. 1993. Removing *giardia* and *cryptosporidium* by slow sand filtration. *JAWWA, Research and Technology*, November 1993, pp. 77–84.

Global Markets Direct 2009. *Global Membrane Market for Water and Wastewater Treatment Forecasts and Analysis to 2015*, 11 November, 2009. http://www.marketresearch.com (accessed on 01/12/2009).

Hinrichsen, D., Robey, B. and Upadhyay, U.D. 1997. Solutions for a water-short world. *Population Reports, Series M, No. 14*. Baltimore: Johns Hopkins School of Public Health, Population Information Program.

Huisman, L. and Wood, W.E. 1974. *Slow Sand Filtration*. Geneva: World Health Organization, pp. 20–46.

Ives, K. 2002. Sedimentation in small community water supplies. Technology, people and partnership, *IRC Technical Paper Series 40*, pp. 313–26.

Lahlou, Z.M. 2003. *Point-of-use/Point-of-entry Systems (POU/POE)*. West Virginia University, US: National Drinking Water Clearinghouse Fact Sheet.

Lampoglia, T.C. and Mendonça, S.R. 2006. *Alcantarillado Condominial. Una estrategia de Saneamiento para Alcanzar los Objetivos del Milenio en el Contexto de los Municipios Saludables*. Lima, Peru: OPS/CEPIS.

Landes, D.S. 1999. *The Wealth and Poverty of Nations*. New York: Norton.

LeChevallier, M.W. and Au, K.K. 2004. *Water Treatment and Pathogen Control: Process Efficiency in Achieving Safe Drinking Water*. Geneva: WHO and IWA.

Lijklema, L., Tyson, J.M. and Lesouef, A. 1993. Interactions between sewers, treatment plants and receiving waters in urban areas: A summary of the Interurba 92 workshop conclusions. *Water Science and Technology*, Vol. 27. No. 12, pp. 1–29.

Lloyd, B.J., Leitner, A.R. and Guganesharajah, K. 2006. *Surveillance for Improvement of Waste Stabilization Ponds. An Investigation and Evaluation Manual*. World Health Organization's Pan American Centre for Sanitary Engineering and Environmental Sciences and Mott MacDonald Ltd, Cambridge, England.

Lyonnaise dês Eaux. 1998. *Alternative Solutions for Water Supply and Sanitation in Areas with Limited Financial Resources*. Paris.

Matsui, S., Henze, M., Ho, G. and Otterpohl, R. 2001. Emerging paradigms in water supply and sanitation. Č. Maksimović and J.A. Tejada-Guibert (eds) *Frontiers in Urban Water Management. Deadlock or Hope*. London: IWA, pp. 229–63.

Meierhofer, R. and Wegelin, M. 2002. *Solar Water Disinfection – A Guide for the Application of SODIS*. Dübendorf, Switzerland: SANDEC/EAWAG.

Metcalf & Eddy, Inc. 2003. *Wastewater Engineering: Treatment and Reuse*. New York: McGraw-Hill.

Moscoso, J. 1998. Acuicultura con aguas residuales tratadas en las lagunas de estabilización de San Juan, Lima, Peru. *Congreso Interamericano de Ingeniería Sanitaria y Ambiental*, 26 (AIDIS), Lima, Peru, 1–5 November 1998.

OPS. 2007. *Tecnologias Alternativas para la Provisión de Servicios de Agua y Saneamiento en Pequeñas Localidades. Memoria del Simposio Internacional*. Programa de Agua y Saneamiento Región America latina y el Caribe, oficina Banco Mundial, Lima, Peru.

Prüss-Üstün, A. and Corvalán, C. 2006. *Preventing Disease through Healthy Environments: Towards an Estimate of the Environmental Burden of Disease*. Geneva: World Health Organization.

Reed, S.C., Crites, R.W. and Middlebrooks, E.J. 1995. *Natural Systems for Waste Management and Treatment*. 2nd edn. New York: McGraw-Hill.

Salazar P.M.L. and Benetti, A.D. 2007. Efecto de las aguas lluvias sobre la eficiencia de reactores UASB. *Conferencia Latinoamericana de Saneamiento – Latinosan*. Cali, Colombia, 12–16 November 2007. http://www.latinosan2007.net/2007/latinosan.htm.

Schmid P., Kohler M., Meierhofer R., Luzi S., and Wegelin M. (2008). Does the reuse of PET bottles during solar water disinfection pose a health risk to the migration of plasticisers and other chemicals into the water? *Water Research*, Vol. 42, No. 20, pp. 5054–5060.

Sharpe, A.J., Quern, P.A. and Currier, R.A. 1994. A consultant's point of view. M.R. Collins and N.D.J. Graham (eds) *Slow Sand Filtration*. AWWA, pp. 37–45.

Skinner, B. and Shaw, R. (date unknown) 58/59. *Household Water Treatment* 1 and 2. Water and Environmental Health at London and Loughborough (WELL), London School of Hygiene & Tropical Medicine (LSHTM) and the Water, Engineering and Development Centre (WEDC), Loughborough University, UK.

Sobsey, M.D. 2002. *Managing Water in the Home: Accelerated Health Gains from Improved Water Supply*. Geneva: World Health Organization.

Solsona, F. and Méndez, J.P. 2002. *Desinfección del Agua*. Lima, Peru: CEPIS/OPS/OMS.

Soman, S.M. 2005. Better access to water in informal settlements. *Sharing, the newsletter of ITDG Sudan*, Issue no. 9, January 2005.

Sommer, B., Marino, A., Solarte, Y., Salas, M.L., Dierolf, C., Valiente, C., Mora, D., Rechsteiner, R., Settler, P., Wirojanagud, W., Ajarmeh, H., Al-Hassan, A. and Wegelin, M. 1997. SODIS – an emerging water treatment process. *J Water SRT – Aqua*, Vol. 46, No. 3, pp. 127–37.

Stanfield, G., Lechevallier, M. and Snozzi, M. 2003. Treatment efficiency. *Assessing Microbial Safety of Drinking Water – Improving Approaches and Methods*, London, UK: IWA, WHO, OECD, pp. 159–78.

Stephenson, D. 2001. Problems of developing countries. Č. Maksimović and J.A. Tejada-Guibert (eds) *Frontiers in Urban Water Management. Deadlock or Hope*. IWA: London, pp. 264–312.

Sulabh International Social Service Organization (SISSO). 2007. *Sulabh Sanitation Movement*. New Delhi, India: CD.

Sutherland, J.P., Folkard, G.K., Mtawali, M.A. and Grant, W.D. 1994. Moringa oleifera as a natural coagulant. *Proc. of the 20th WEDC Conference*. Colombo, Sri Lanka.

United Nations Environment Programme (UNEP). 1998. *Sourcebook of Alternative Technologies for Freshwater Augmentation in Latin America and the Caribbean*.

United Nations Environment Programme (UNEP). 2002a. *Rainwater Harvesting and Utilisation. An Environmentally Sound Approach for Sustainable Urban Water Management: An Introductory Guide for Decision-Makers.* http://www.unep.or.jp/ietc/Publications/Urban/UrbanEnv-2/index.asp.

United Nations Environment Programme (UNEP). International Environmental Technology Centre (IETC). 2002b. *Environmentally Sound Technologies in Wastewater Treatment for the Implementation of the UNEP Global Programme of Action (GPA) 'Guidance on Municipal Wastewater'.* http://www.unep.or.jp/Ietc/Publications/index_pub.asp.

United Nations Environment Programme (UNEP). 2002c. State of the environment and policy perspective: 1972–2002. *Global Environment Outlook 3.* Chapter 2. Stevenage, UK: Earthprint.

United Nations Environment Programme (UNEP). 2004. *Saving Water through Sustainable Consumption and Production: A Strategy for Increasing Resource Use Efficiency.* Paris: Division of Technology Industry & Economics (DTIE)/UNEP. http://www.unepie.org/pc/cp/reportspdf/GMEFwater.pdf.jp/Ietc/Publications/index_pub.asp.

United Nations Environment Programme (UNEP). Stockholm Environment Institute (SEI) 2009. *Rainwater harvesting: a lifeline for human well-being.* Paris: UNEP.

US-EPA. 1999 *Alternative Disinfectants and Oxidants*, EPA Guidance Manual, United States Environmental Protection Agency, US, pp. 3.1–3.52.

US-EPA. 2006. *Emergency Disinfection of Drinking Water.* EPA 816-F-06-027, Office of Water 4606-M, United States Environmental Protection Agency, US.

Van Haandel, A.C. and Lettinga, G. 1994. *Anaerobic Sewage Treatment: A Practical Guide for Regions with a Hot Climate.* Chichester: Wiley.

Veenstra, S., Alaerts, G.J. and Bijlsma, M. 1997. Technology selection. R. Helmer and I. Hespanhol (eds) *Water Pollution Control. A Guide to the Use of Water Quality Management Principles.* London, WHO/UNEP, pp. 45–85.

Water and Sanitation Program (WSP). 2007. *La Ciudad y el Saneamiento. Sistemas condominiales: Un Enfoque Diferente para los Desagües Sanitarios Urbanos.* Programa de Agua y Saneamiento para America Latina y el Caribe, Lima, Peru.

Whittington, D., Lauria, D.T. and Mu, X. 1989. Paying for urban services – A study of water vending and willingness to pay for water in Onitsha, Nigeria. *The World Bank Policy, Planning and Research Staff,* Infrastructure and Urban Development Department, Report INU 40.

WHO. 2002. *The World Health Report. Reducing Risks, Promoting Healthy Life.* Geneva: World Health Organization.

WHO. 2004. *Guidelines for Drinking Water Quality.* Vol. 1: Recommendation. 3rd edn. Geneva: World Health Organization.

WHO. 2006a. *Guidelines for the Safe Use of Wastewater, Excreta and Greywater: Wastewater Use in Agriculture.* Vol. 2. Geneva: World Health Organization.

WHO. 2006b. *Guidelines for the Safe Use of Wastewater, Excreta and Greywater Wastewater and Excreta Use in Aquaculture.* Vol. 3. Geneva: World Health Organization.

WHO and UNICEF. 2000. *The Global Water Supply and Sanitation Assessment 2000.* Geneva: World Health Organization.

WHO and UNICEF. 2004. *Meeting the MDG Drinking Water and Sanitation Target: A Mid-term Assessment of Progress.* Geneva: World Health Organization.

Chapter 3

Stormwater management in the humid tropics

Joel Avruch Goldenfum[1], Carlos Eduardo Morelli Tucci[1,2], and André Luiz Lopes da Silveira[1]

[1]Institute of Hydraulic Research, Federal University of Rio Grande do Sul, Porto Alegre, Brazil
[2]FEEVALE University Center, Novo Hamburgo, Brazil

Developing countries constitute the great majority of the tropics. Technical tools and principles of urban drainage control that are well-known in developed countries are not always successfully applied there. Consequently, stormwater management measures in the humid tropics should be adapted to both climatic and socioeconomic conditions. As such, the principal elements of stormwater management in the tropics are: climate, socioeconomic factors and institutional aspects.

This chapter presents urban drainage management. It describes the main principles as well as related factors, such as climate conditions, socioeconomic factors, structural control measures, urban drainage policies, and institutional aspects, taking into account the difficulties asscoiated with the social, economic and environmental conditions that characterize the majority of countries in the tropics.

3.1 URBANIZATION AND HYDROLOGY IN THE HUMID TROPICS

Urban drainage serves to reduce the risk of flooding and inconvenience due to surface water ponding, alleviate health hazards, and improve the aesthetics of urban areas. This service is provided by two types of sewers, separate or combined. The separate system consists of two independent systems of sewers: storm sewers conveying surface runoff (stormwater) and sanitary sewers conveying municipal sewage. In the combined system, both sources are conveyed in a single system, which may overflow in wet weather – the escaping flow being termed a combined sewer overflow (CSO) (Marsalek et al., 2008).

In the tropics the solutions usually applied to urban drainage problems aim to rapidly remove excess stormwater by transporting it away, normally transferring the problem downstream.

Urban drainage control measures, however, must be approached from an environmental standpoint, and should be based upon appropriate management of urban impacts on the hydrologic environment. This approach is complex and incorporates engineering, sanitary, ecological, legal and economic technical perspectives, as well as being closely related to the design and management of urban spaces. The hydrological cycle is a key element in defining urban sanitation and drainage.

The environmental (also called alternative or compensatory) approach is the complete opposite of the traditional approach, concerned as it is with the maintenance and recovery of environments with a view to maintaining a healthy urban area both inside and outside the city. The drainage and sewage treatment equipment should characterize a sanitation system that is part of an overall organization of urban spaces that values, preserves and recovers watercourses. This leads to the notion of the self-sustainability of cities and towns, in particular regarding the internal and external environment. The sanitary and drainage equipment helps preserve the water quality of watercourses, both inside and out of the city, and thus contributes to ensuring the city's sustainability (Sangaré, 1998). Stormwater and sanitary drainage systems and their treatments provide the city with autonomy, thereby sparing its receiving bodies – water courses and the water table. Consequently, the autonomy of the latter is favoured in the city, because their recovery and conservation encourage the usual and permanent biological development of species that naturally inhabit the area.

Part of the urban drainage problems in developing countries results from the neglect of stormwater drainage systems in housing and other developments. The 'divorce' between urban growth and wastewater systems has caused critical problems in many cities related to internal inundation, prolonged flooding and increasing levels of pollution in receiving bodies. This has already happened in temperate climates but the climatic differences in the humid tropics only tend to worsen the consequences of anthropogenic occupation over uncontrolled urban runoff.

3.1.1 Unregulated developments

Urban development in developing country cities is rapid and unpredictable and often ignores city regulations. As a consequence, private landowners often develop land without the necessary infrastructure. This land tends to become inhabited by low-income communities. Neither the municipality nor the population have sufficient funds to provide basic water, sanitation or drainage infrastructure, and there is a general lack of capacity to manage urban stormwater using principles of sustainability. Local authorities tend to respond reactively to flood events, but rarely follow long-term strategic plans incorporating a pro-active, preventative approach.

The poorest communities often end up living precariously in areas like low-lying lands adjacent to watercourses or hillside slopes, which are at the greatest risk from natural flooding events (such as in Bangkok, Bombay, Guayaquil, Lagos, Monrovia, Port Moresby and Recife) or landslides (such as in Caracas, Guatemala City, La Paz, Rio de Janeiro and Salvador) (WHO, 1988).

Lack of appropriate refuse collection decreases the capacity of the urban drainage network, which can lead to an almost total loss of function, as in the case of some African countries (Tokun, 1983; Desbordes and Servat, 1988).

3.2 CLIMATIC FACTORS IN THE HUMID TROPICS

The two main types of urban floods are:

- *local floods* generated within the urban environment itself, which are routed through streets, stormwater sewers and creeks, and
- *riparian floods* which affect urban areas but are caused by floods in rivers or lakes close to the city.

Climatic conditions in the humid tropics expose cities and towns to more frequent floods of both types, in which associated damages are aggravated by socioeconomic factors. The most important climatic factor in the humid tropics is undoubtedly higher rainfall, both in terms of volume and maximum intensities. The intensities of storms lasting less than one hour are significantly higher than those in temperate countries, and this greatly increases peak flows and volumes flowing into drainage structures.

Hydraulic structures must therefore have higher capacities for conveyance and storage of floodwaters, and are consequently more expensive. The higher intensities favour a greater capacity for erosion of solid transport during the first flush period,[1] burdening structures that are designed to trap solids. In addition, the larger mean volumes of precipitation and higher rainfall intensities lead to greater challenges for management of urban runoff since:

i)   larger volumes must be routed somewhere
ii) the mean transported load of solids is larger (because runoff continues for a longer time)
iii) there is less time to maintain drainage structures during periods of dry weather, and
iv) there is more time for disease vectors and water-related diseases to develop.

Higher rainfall means costly structural measures that are more difficult to implement. Non-structural measures may also be affected by the rapid rise in floodwater level, greater peaks of runoff and longer flood durations compared with temperate climate. Consequently, warning times are shorter, more areas are flooded with greater risk, and it is more expensive to adapt residential or commercial buildings to withstand floods.

In addition, the higher ambient temperatures in the humid tropics favour the proliferation of insect and animal vectors that carry diseases. This, together with the prevalence of accumulated water in drainage structures due to high rainfall, may increase the risk of disease outbreaks. There are limitations related to the use of storage structures, and their operation requires consideration of the dangers relating to infectious agents such as mosquitoes.

3.3 SOCIOECONOMIC AND INSTITUTIONAL ASPECTS

Problems related to urban stormwater management in the humid tropics tend to occur as a result of the combined complex and competing economic and social forces observed in cities in low-income countries, and the lack of capacity within local governmental authorities to deal with these complexities.

Uncontrolled urban developments, whether legal or illegal, occupy river basins and floodplains, increase rates of soil imperviousness, encroach on natural watercourses (forcing the construction of open or closed canals), and cause increased runoff, which

---

[1] First flush is the initial surface runoff of a rainstorm. During this phase, water pollution entering storm drains in areas with high proportions of impervious surfaces is typically more concentrated compared to the remainder of the storm. Consequently, these high concentrations of urban runoff result in high levels of pollutants discharged from storm sewers to surface waters.

consequently requires larger storm drainage systems as compared to cities in developed countries in temperate climates. In addition, high urban population densities, including not only urban slums but also many middle-class neighbourhoods, reduce the availability of open spaces, which could be used as part of an integrated stormwater management system.

Socioeconomic, rather than climatic, factors play a relevant role in inhibiting the introduction of environmental practices in urban drainage. Riparian areas of creeks are often occupied to the limit and urban occupation is dense with high rates of imperviousness. This situation is visible both in middle-class neighbourhoods and slum areas since rates of imperviousness, modification of creeks – which are frequently straightened or canalized – and occupancy of riparian areas occur in both cases.

According to Ruiter (1990), Asian cities demonstrate a lack of institutional organization and clear allocation of responsibilities, which is compounded by inadequate skills and resources for project management, urban land-use planning and enforcement, as well as aspects related to technical and non-structural planning.

The lack of flood control management related to urban development is a common situation in cities of developing countries. There are significant constraints on the development and successful implementation of sustainable stormwater drainage plans which relate to various institutional aspects. In some cities, appropriate legislation exists, but is not implemented in a great number of situations for numerous reasons. These may be attributed to:

i) poorly-defined roles and responsibilities for different aspects of stormwater management amongst different urban institutions
ii) lack of regulation enforcement due to weak county institutions or small number of personnel
iii) low price of real estate and high cost of implementing regulation requirements
iv) ineffective laws and regulatory instruments at the municipal level related to quantitative and qualitative aspects of urban drainage
v) a lack of sufficient finances and other resources due to a lack of political will to invest in stormwater management, and a lack of willingness to pay in the long term for services which may not be utilized on a regular basis, and
vi) a lack of qualified professionals with a sufficient level of understanding of flood-control measures

Since flood control and urban drainage are rarely charged as a service, city authorities seldom have funds to deal with the problem. The majority of funds come from emergency reserves, and flood management is often undertaken through federal or state organizations that have much broader goals than the specific conditions of individual cities.

3.4 URBAN DRAINAGE STRUCTURAL CONTROL MEASURES

In the majority of situations, the types of hydraulic structures and ancillary equipment developed in temperate climates are applied in the humid tropics with little or no modification. In the next section of the chapter, the most frequently-used structures are presented, and their advantages, disadvantages and applicability to climatic and socioeconomic conditions in developing countries in the humid tropics are discussed (Silveira et al., 2001).

Flow control structures are divided into three basic types:

i)   storage,
ii) infiltration, and
iii) floodplain and channel management.

Storage-type structures can be subdivided into off-site and on-site. The main off-site type works are detention ponds and retention ponds, which may also include the development and control of natural lakes.

On-site structures can form part of a chain of smaller reservoirs and hydraulic devices designed to control runoff in parks, residential units, parking lots, housing estates, schoolyards and so on. The infiltration-type structures include pervious pavements, infiltration trenches, ponds and inlets.

However, conventional drainage structures are also needed as part of a comprehensive stormwater management strategy, which needs to integrate both transport and detention structures. But these present different levels of performance in relation to drainage efficiency and pollutant removal. Some of these structures are described below and restrictions to their use in humid tropics are also indicated.

3.4.1 Storage-type devices

Structural measures compatible with the objectives of sustainable urban drainage include detention and other attenuation structures. However, particular care is required for their design as the operational performance of these structures in humid tropics is less well-known than conventional structures for rapid drainage of floodwaters.

Detention ponds

Detention ponds are reservoirs that are dry during low-flow periods and whose purpose is to attenuate peaks of surface runoff and thus release flood volumes gradually. The ponds can be excavated into the ground or formed by construction of a small earth or concrete dam, potentially using natural depressions in the terrain.

To operate appropriately, they require the installation of devices such as a settling basin and racks at the inlet to impede the entry of sediments and refuse. At the outflow, an emergency overflow is needed to bypass flows greater than the design flow.

Reservoir volumes are sized in temperate climates using different methods, but these are normally the equivalent of surface runoff volume generated in a catchment as a result of a 24-hour duration storm with a return period of one to two years. The runoff coefficient is generally taken as 0.50 or 0.60. Another form of sizing is the imposition of a maximum discharge at the outlet, often similar to those under natural conditions with a one or two-year return period. Once the necessary storage volume is known, economical criteria can be used to select the structural dimensions.

In order to ensure their pollution removal role, outlet structure devices must be sized so as to detain part of peak stormwater flows, long enough to settle the solids.

The mean stormwater flows are discharged within 12 to 24 hours, and the outlet structure is sized to release the design volume in double or treble the time (Urbonas and Stahre, 1993).

In the humid tropics the highest storm intensities may force engineers to reduce the design return period, in order to have an economically feasible structure. This evidently reduces the efficiency of these structures compared with temperate climates. In addition, the greater frequency of rainfall may render the concept of detention unfeasible since the reservoir may never be sufficiently empty to act as a storage device.

In order to avoid connections from foul sewers to the storm drainage system, it is important that systems for the appropriate disposal of domestic sewage are installed upstream of detention basins.

Another important consideration in the humid tropics is the potential proliferation of mosquitoes in detention basins. It is important to know the timing of the reproductive cycle of the local species, as this may be used to determine hydraulic detention times. As long as these are less than the local reproductive cycle, it is difficult for mosquitoes to breed. However, the frequency of rainfall in the humid tropics means that detention tanks are rarely empty, and these can become a loci for mosquito breeding.

In addition to climatic characteristics that favour the proliferation of disease vectors, detention basins in humid climates are likely to be less efficient and have a more limited use compared with those in temperate climates. This is due to a lack of regulatory controls to ensure that polluted wastewater does not inflow into the tank, as well as the lack of control of urban developments on areas suitable for locating detention ponds.

Retention ponds

Retention ponds are primarily designed to improve the quality of water from stormwater flows, but are often employed as flood control devices. They are designed not dry out between rainfall events, thus retaining water permanently as a part of their volume. The basic design parameter is the residence time, generally between two to four weeks.

Water quality improvement occurs through settling, and moderate to high rates of pollutant removal are achieved if the permanent water volume in the reservoir is between approximately 30 and 60 mm of runoff per hectare (0.5 to 1.0 inch per acre) of stormwater flows of impervious surfaces.

In temperate climates, implementation is restricted by the difficulty of maintaining a permanent volume of water if the contributing area is less than 4 ha (10 acres), and the drainage area/water surface ratio is less than 6:1. This difficulty may be minimized in the humid tropics due to higher rainfall.

As with detention basins, there are constraints on the implementation of retention ponds due to their space requirements and the problems related to contamination from inflows from foul wastewater and consequential environmental health problems. Contamination by sediments, refuse and sanitary sewage may cause much more damaging effects than in temperate climates, notably due to the proliferation of mosquitoes that act as disease vectors for malaria, dengue and yellow fever, and rats which transmit leptospirosis.

Thus, as with detention tanks, sediment and refuse trapping structures and sanitary sewage networks must be carefully isolated upstream. Specifically with regards to mosquitoes, a special stormwater and mosquito abatement scheme must be set up, possibly with scheduled emptying during critical periods of mosquito development which present a danger to public health.

On-site detention(OSD)

These structures refer to small reservoirs built to attenuate stormwater flows produced in residential and commercial urban lots of up to $600 \, m^2$. These generally consist of simple concrete box-shaped structures with an orifice-type outlet device which restricts discharge up to a specified design value.

OSD may be retrofitted in existing buildings without the need for major architectural or urban adaptation, but in most cases these structures are installed by private developers to ensure that new developments fulfil a legal restriction on stormwater runoff production, generally specified in the form of a permissible site discharge (PSD). PSDs must be evaluated in the context of the catchment, so that the flows from each site will not violate the limit established for it.

OSD structures for lots smaller than $600 \, m^2$ are not usual in the US and Europe, but have been widely applied in Sydney (Australia) with typical reservoir volumes ranging from 200 to $500 \, m^3$ per site hectare (Nicholas, 1995). In the humid tropics, OSD could become a useful alternative to other forms of off-site attenuation described above, since their installation in dense urban areas is not an obstacle. Efforts to control runoff problems in urbanizing slums might utilize OSD structures, which could, initially involve modification to existing water tanks.

One disadvantage of higher rainfall is the requirement for larger reservoir volumes to cope with a rainfall event, as compared to those necessary in a temperate climate for a risk of the same occurrence. From the sanitary standpoint, special care is required in designing and cleaning OSDs to avoid accumulation of water and dirt and to reduce the danger of development of tropical disease vectors.

3.4.2 Infiltration-type devices

Infiltration trenches

Infiltration trenches are devices that promote the infiltration of runoff into the natural soil. In the humid tropics, the main potential uses for infiltration trenches are outdoor parking lots of residential buildings and commercial businesses such as supermarkets and shopping malls.

Trenches are excavated in the soil and filled with uniformly crushed stone. The walls and the top are lined with a pervious filter fabric to avoid sediment penetration. The infiltration rate must not be too low so as to make emptying times too long, nor so high as to result in contamination of the water table due to lack of adsorption of pollutants onto the filter media because of rapid movement of water through the soil.

The main function of the infiltration trench is to reduce peak runoff discharge and promote aquifer recharging. However, another important function is to assist the treatment of runoff by means of infiltration into the soil. However, these devices cannot function effectively where sediment concentrations in runoff are high, since this compromises long-term permeability of the filter media and underlying soil. Therefore, upstream sediment removal structures, in the form of settling basins or grass filter strips, must be installed.

In the site layout, infiltration trenches should be long and narrow, and placed upstream from the conventional storm system. Their use is suggested for parking lot perimeters, but is not recommended for industrial or commercial areas that store chemicals, pesticides and oil derivatives, or close to water well uptake sites.

Grass filter strips

Grass filter strips should be located upstream from the drainage system and are designed to attenuate and partially infiltrate runoff from urban impervious surfaces (parking lots and other surfaces), although their use may be associated with other situations. The linear character of grass filter strips allows for very flexible spatial arrangements (Schueler, 1987). The size of a grass filter strip depends on the particulate trapping, efficiency desired and the slope (Mecklenburg, 1996).

Filter strips significantly reduce the speed of runoff, but do not greatly reduce peak runoff to pre-development levels. Their main benefit is the removal of polluting particles such as fine sediments, organic matter and metal traces (Schueler, 1987), but grass filter strips are not effective for the removal of soluble pollutants (Mecklenburg, 1996).

Infiltration capacity can be evaluated by means of infiltration trials in experimental studies. A grass filter strip should be as flat as possible and the grass should be dense to avoid erosion and worse effects of concentrated runoff. This aspect should receive particular attention in the humid tropics, as intense storms have great kinetic energy, and therefore great erosive potential.

Grassed swales

Grassed swales are grassed channels in which stormwater runoff is slowed down and partially infiltrated along their course, the excess being sent into the conventional storm drainage network. They are appropriate for residential lots where they replace conventional canalized sewerage, reducing both runoff and acceleration of excess stormwater.

Like grass filter strips, the effects expected from grassed swales (stormwater or water quality benefits) are only significant for slopes less than 5% (Schueler, 1987). In the humid tropics greater difficulties are expected from the operation of swales because of very heavy rainfall, which may rapidly saturate the soil and promote a larger volume and faster runoff, thus cancelling out their effect. For less intense, but frequent, rainfall in the humid tropics, the constant presence of water in the swales may cause discomfort and be dangerous from a sanitary standpoint.

Pervious pavements

Pervious pavements are infiltration devices where runoff is routed through a permeable surface, into a rock reservoir located under the soil surface. As long as the soil is not saturated, pavements act as peak flow controls, reduce total runoff volume, promote groundwater recharge and also contribute towards pollutant removal.

There are three types of pervious pavements: porous asphalt pavements, porous concrete pavements and garden blocks. The main difference with conventional pavements is the lack of fine sand in the structure. Instead, pervious pavements are usually located over a granular base and geotextile filters are used to avoid fine sand migration into the reservoir (Urbonas and Stahre, 1993).

Light honeycomb structures of heat-formed plastic materials are also applied, as an alternative to underground reservoirs. Due to the bigger sediment yield in the humid tropics, clogging of the top surface of pervious pavements can be a major problem, drastically reducing their lifetime. This problem is increased when there is lack of proper maintenance. The use of pervious pavements in the humid tropics is generally limited to pathways and shopping centre parking areas, where the maintenance can be more easily ensured, reducing the danger of clogging.

Infiltration ponds

Also called infiltration basins, these are isolated areas on the terrain, specifically designed to infiltrate excess stormwater runoff, which is routed into the soil. Although similar to detention basins, the main difference is that there is no outlet for deliberate emptying. However, for safety reasons an emergency hydraulic overflow is included, and a backup under-drain to preserve the base.

Infiltration basins are appropriate for places with pervious soils and a deep water table. Schueler (1987) states that the main advantages of infiltration basins are that they preserve the local water budget, control peak discharges, can be used as sediment basins during the site construction phase, and are affordable.

The disadvantages are that they cannot be applied in soils that are not very pervious, the need for frequent maintenance, the presence of possible bad odours, the potential for mosquito development, and the constantly muddy bottom, all of which make their use as recreation areas difficult. In the humid tropics these disadvantages can lead to decisions not to use infiltration ponds, since storms are more intense and frequent than in temperate climates.

Schueler (1987) mentions that an infiltration basin can control a drainage area of up to 20 ha (50 acres). The volume destined for infiltration is created by excavation or an embankment. The bottom of the basin should be flat and lined with dense grass. Additional temporary detention for a two-year and/or ten-year design storm can be included in the form of a combined infiltration/detention basin.

Infiltration inlets

These are draining structures that replace inlets for conventional stormwaters (gully-holes). They constitute linear structures similar to infiltration trenches, the difference being that the bottom is isolated with geotextile film. The runoff moves through a bed of round stones, which attenuate the flow of runoff into the main stormwater drainage system. There are restrictions to their use in the humid tropics due to the larger volumes that flow into them, which mean that the infiltration inlets need to be proportionally larger.

3.4.3 Control of urban solids

Legal and illegal domestic sewer connections to stormwater drainage systems mean that ancillary structures, such as retention basins, become pollutant traps, which are washed out during high flows and lead to water contamination.

The lack of control and inspection of construction sites also results in high sediment concentrations in runoff. Refuse is another problem, caused either by the trend towards higher per capita production or lack of environmental education, which leads residents to use urban creeks as disposal sites. These factors, combined with ineffective systems for drain cleaning and solid waste collection, mean that high runoff flows become the *de facto* mechanism for removal of solids from urban surfaces.

The volume of solid waste is often so large that the conventional structural solutions of cities in developed countries do not function effectively. New approaches must therefore be sought in developing areas in the humid tropics.

Litter traps

These are very useful structures in developing countries where systems for solid waste collection are inadequate and ineffective. The amount of refuse on the streets that has not been properly stored or collected can result in a significant quantity being carried off by storms into the drainage system and creeks. In the humid tropics this condition is worsened by the significant transport capacity associated with high intensities of precipitation and frequent rainy days.

The situation is worse on the outskirts of cities in developing countries, where urban equipment and municipal services are even more precarious, and the population discharges their refuse into urban creeks. Even where a drainage system has been implemented, gutters, gully holes and buried storm drains in cities are commonly used as places to dump and remove refuse. Consequently, large volumes of refuse end up in the creeks and other storm drainage receiving bodies, even when they do not completely block the formal drainage.

3.4.4 Floodplain and channel management

Forested buffer strip

A forested buffer strip is an area of forest close to streams, rivers and other water bodies, designed to remove pollutants associated with urban stormwater runoff. They also reduce erosion, reduce water temperatures (by shading) and ameliorate the wildlife habitat (Mecklenburg, 1996). Due to its characteristics, it can be considered a post-construction control device.

In major cities in tropical countries, the land adjacent to water bodies is generally occupied by the poorest segment of the population, and their removal in order to build a forested buffer strip can constitute a social problem. After their implantation, these areas can be popular recreation spots, but are frequently illegally occupied and can become dangerous areas, with the presence of muggers and drug addicts.

Level spreader

According to Mecklenburg (1996), a level spreader is a weir used to stop the formation of gullies. Its top is built at the same elevation as the surrounding ground, so that it can convert concentrated flow to sheet flow. Level spreaders can increase infiltration and groundwater recharge, preserve natural vegetation, and reduce disturbance of the riparian corridor and stream channel.

These structures can provide a good solution for the prevention and control of gully formation in areas where the natural topography does not concentrate flows immediately below the point of discharge. Their use should be avoided in highly erosive areas.

3.5 URBAN DRAINAGE AND URBAN MASTER PLANNING

3.5.1 Overview of the planning process

The problems of urban development in the tropics are not only the result of climatic conditions; other important factors are density of occupation, lack of regulations and law enforcement – mainly due to weak institutional capacity – and illegal occupation of public or private areas.

The integration of urban planning and stormwater management should ideally be based upon preventative action, since the costs are lower than those of corrective action, and it is technically simpler to implement. Stormwater plans need to be developed within the context of the local environmental, physical and socioeconomic situation. Each problem has its own solution and each situation requires a different approach to developing urban drainage and flood control plans. Those responsible for developing plans need to understand local urbanization pressures and the behaviour of water in the basin, as well as the important role the water cycle plays in relation to public and environmental health.

The development of urban flow controls needs to be based primarily upon hydrological sub-basins, but recognition of political and administrational boundaries is critical to ensuring the creation of effective institutional and regulatory systems. These systems should cover all the areas described in the Urban Master Plan; they also need to address any transboundary matters and broker agreements with the relevant institutional stakeholders. Such agreements are important because it is impossible to enforce regulation if only one side of the basin is regulated.

3.5.2 Strategies for urban stormwater management

Experiences in flood control in many countries have led to the following principles, which underlie strategies for urban stormwater management:

- Flood-control evaluation should be undertaken in the basin, and not only in specific flow sections.
- Urban scenarios should take into account future city developments.
- Flood-control measures should prioritize source control measures, and should not transfer the flood impact to downstream reaches.
- The pollutant impact caused by urban runoff and other issues related to urban drainage water quality should be reduced.

- More emphasis should be given to non-structural measures for floodplain control such as flood zoning, insurance and real-time flood forecasting.
- Management of control starts with the Urban Drainage Master Plan, which is implemented in the municipality.
- Public participation in urban drainage management should be increased.
- The development of urban drainage should be based on cost recovery investments.

Based upon these principles, we can categorize the main strategies for urban stormwater management plans into *structural* and *non-structural* strategies, as summarized below.

Structural strategies

1) Surface runoff should be distributed throughout the catchment as much as possible, making the best available use of open spaces, decentralized storage and the hydraulic capacity of both natural and constructed drainage channels.
2) When there is a need to install storm drains and/or increase existing channel capacity, the plan needs to be designed so as to minimize downstream impacts. To ensure flood drain capacity is not lost over time, specific attention must be paid towards solid waste management and the instigation of strategies (structural and non-structural) to reduce the influx of solids into drains.

Non-structural strategies

1) Developments on floodplains need to be controlled through the enforcement of land zoning strategies according to the level of flood risk in each area.
2) Provision must also been made for flood preparedness and response strategies using non-structural measures and relief measures for infrequent but high-impact, large-scale floods.

In low and middle-income countries, there are many uncertainties related to investment in urban stormwater management and flood control due to uncontrolled urbanization. Economic studies can be undertaken to assess risks and the costs and damages versus the benefits using different future scenarios of urban development and flood return periods.

3.5.3 Institutional considerations

Without effective regulation and legal enforcement, the installation of source control technologies to prevent expected increase in runoff from new urban developments is severely limited. Delays in planning and the effective implementation of non-structural measures for the control of development on open areas invariably result in increased costs for future remedial actions.

In developing countries, part of city expansion is based on illegal developments, while most of the city's densification (approval of construction in existing urban infrastructure) follows national administrative regulation, as the resulting real estate has greater market value. This is not the best point at which to regulate drainage control, but for city areas where downstream drainage is a problem, source control at this stage of development can be an important management option. Regulation can be achieved through permeable areas or on-site detention (see below).

There are many small counties where local technical, economical and financial capacities are too small to cope with this type of control, and support is needed from state or national/federal agencies. The main flood control and urban drainage actions that can be developed at management level by public organizations are recommended below.

Federal level

The main measures and preventive programmes at the federal level are the following:

- Communities should be supported with non-refundable funds only when preventive flood control programmes are in place and an Urban Drainage Master Plan has been developed. For those without preventive programmes, support funds should be provided as a loan.
- Avoid the construction of public buildings, hospitals, schools and other such buildings in flood-risk areas with bank loan funds. When there is no other option, these should be built to a high flood-security standard.
- Whenever possible create and promote a flood insurance system.
- Support preventive programmes and nforms of non-structural flood control such as: flood zoning, insurance, flood forecasting and regulation of urban drainage.

State level

For countries where the state is a political agency, it can provide support to counties/regions on preventive measures and help solve conflicts among more than one county/region. Types of state support may include the following:

- Retain a state technical team to provide support to counties/regions on flood control and urban drainage, in particular, to advise on preventive programmes of planning, regulation, civil defence and environmental issues.
- Establish standards and state measures for minimization of flood impacts.
- Develop a monitoring system for hydrologic data, flood forecasting and water quality of nearby rivers.
- Develop economic and tax mechanisms to improve flood and water-quality control.
- Evaluate flood losses and the results of preventive programmes.
- Develop Urban Drainage Plans for Metropolitan Regions that cover many counties/regions. This is necessary in order to manage conflicts created by urban drainage impacts in the same urban basin.

Municipal level

From an operational perspective, the management of urban drainage and flood occupation is usually undertaken at the municipal level, including both non-structural and source control measures. It is at the municipal level that the Urban Drainage Master Plan is also developed and where laws and regulation are implemented.

Some of the main aspects which should be included in the plan are as follows:

- hydrologic, hydraulic and environmental standards for planning and design described in a standard manual
- development of efficient processes for law enforcement and maintenance of public services
- capacity-building for technical staff, and
- public participation.

These basic elements should be adapted to the conditions of each city. The following section presents some of the main concepts of Urban Drainage Master Plans and Participatory Master Plans.

3.5.4 Participatory planning processes

Successful source control strategies are dependent upon effective regulatory instruments together with institutions that have the capability to enforce these regulations. However, due to the great differences in social and economic situations, there is a strong need for public participation to ensure support from local residents. In addition, when new political administrations assume office, many projects are interrupted or terminated. Increased public participation gives long-term planning a greater chance of surviving political changes (Goldenfum et al., 2007).

The adoption of compensatory devices often faces opposition from the population, designers and public managers, mainly because of lack of knowledge of infiltration and storage devices, and also due to natural resistance to new concepts and ideas (Baptista et al., 2005). It is necessary to implement an educational process directed at all sectors involved in planning, installation and maintainance of urban water systems.

The development of a Participatory Urban Master Plan can be an opportunity to discuss and diffuse knowledge on sustainable urban water management to different sectors of society. The participation of urban water professionals in the development of these Master Plans can allow for the spread of sustainable urban water management concepts.

Public participation can be achieved through the organization of community-level consultation and decision-making sessions, which explain the impacts and benefits of the Plan and associated projects during their development, and receive inputs from the local population.

Although the conventional approach based on the rapid drainage of excess runoff still prevails, some Brazilian cities (such as Belo Horizonte, Porto Alegre, Guarulhos, Curitiba and Caxias do Sul) have already begun incorporating sustainable urban water management concepts, based upon principles of public consultation.

In Brazil, the Statute of the City (Brazilian Law 10.257, 2001) establishes standards to regulate urban propriety usage, aiming to achieve public welfare and environmental balance. According to this law, the Urban Master Plan is an essential part of the Municipal Planning Process and is the fundamental instrument for urban development.

The Statute of the City also foresees the development of Participatory Master Plans, which would establish a new forum to discuss urban planning aspects with different sectors of society.

During the development of the Participatory Master Plan, public audiences and meetings are organized, during which the technical team present their analysis and diagnostic of present conditions as well as their views on development, and local society present their own perceptions in response. Additional meetings with particular economic and public local groups to discuss the pertinent elements of each interest group may be required:

• thematic meetings – to discuss specific aspects related to each of the strategies from the Strategic Plan
• 'territorial' meetings – to discuss local questions from each municipal region
• proposal Presentation Public Audience – to detail and discuss the Proposals Document, which gathers all conclusions from the previous meetings and forms the basis for the draft Urban Master Plan, and
• the Public Final Conference, to present and discuss the Final Proposal.

3.5.5 Urban drainage master plans

Usually, an Urban Plan is based upon goals and objectives related to the well-being of the population and environmental conservation. In Urban Drainage the main goals are as follows:

• Urban drainage planning aims mainly to distribute volume allocation in time and space in the urban basin, based on urban spaces, hydraulic network and environmental conditions.
• It aims to control the occupation of floodplain areas through regulation and other non-structural measures.
• It aims to provide revention and relief measures for low-frequency floods.

Urban flow control is developed by sub-basins and is regulated by modules defined by political city divisions, but have technical restrictions related to basin flow. The main flood control policies may be summarized based on drainage system, as follows.

In the main drainage system: reserve urban space for detention or create linear parks within the river boundaries for damping the flood peak, sediment and trash detention and water quality improvement. Since part of upstream urbanization is uncontrolled due to lack of law enforcement (as described earlier), this policy will avoid urban drainage impact transference to downstream reaches. Instead of being distributed in conduits or along rivers and channels, refuse and sediments are retained in specific places for maintenance purposes. This may not be the optimal solution for certain situations, and should be evaluated based on local conditions. It is important to evaluate the sewage and refuse systems within this framework, since they should be mutually compatible.

When the selected solution for flood control is stormsewers or increased channel capacity, the plan or design has to control their downstream impacts.

In the floodplain: use non-structural measures such as flood area zoning of low and high-risk areas, constructions standards and overall regulation related to this control. Some common recommendations include the following:

- Public flood areas whose value has increased due to urban development pressure should be developed with urban infrastructure, such as parks, sports fields, playgrounds, and so on. If left undeveloped, these areas could be invaded.
- The same practice should be followed when a private area is bought for relocation. Public infrastructure should be constructed immediately following relocation, as other families may move to the same area.
- Education of neighbourhood families concerning flood impacts can usually preserve an empty space.
- Tax benefits should be improved or alternatives uses found for private areas.

The Plan normally comprises the following studies:

- flow control in the main drainage of the city
- proposed regulation for developments in new areas inside the city boundaries, and
- development of other institutional aspects such as: organization and personnel, capacity-building, public participation, and others.

The Urban Drainage Plan usually has the following steps:

1. Development of solutions for current development scenario based on public investment (since floods are an existing problem).
2. Development of regulations related to urban drainage, taking into account social and economic conditions. The best form of regulation increases public participation. Basic aspects of this type of regulation are: i) the new development has to keep peak discharge equal to or below the pre-development scenario, and ii) limits for impervious areas.
3. Planning the development of public green areas for parks which could be used as detention for flood rise not held by regulation. Such recreational sites also improve the environment. This is a prevention process since these areas may be used for future detention. If these areas are used by natural drainage they could be also help control drainage impacts from urbanization.

The three main products of the Urban Drainage Master Plan are the Action Plan, Regulation and the Urban Drainage Manual. These are described below.

Action plan: this should include the main actions needed to implement the measures proposed in the study for critical areas. It usually includes an emergency plan for serious problems which require a quick solution. Medium and long-term actions have to be developed over time in order to implement all public actions relating to urban drainage.

Regulation: this forms part of the non-structural measures for space control. It is established by county/regional legislation or is included within the city building code. The non-structural public measures in the plan should focus on:

- urban drainage regulation for urban densification and new developments
- identification of green public areas for prevention
- development of public education and participation, and
- training of engineers and architects.

Urban drainage manual: the urban drainage manual is used to advise engineers on accepted city restrictions and procedures in urban drainage design.

The manual should present the main conceptual elements of the Urban Drainage Master Plan, such as: source control (holding peak discharge increase without transferring its impact downstream), improving water quality, and taking into account future urban scenarios.

Regulation related to drainage should be clearly identified in the manual, including: i) flow limits for selected return periods, ii) detention and impermeable area limits, iii) urban densification, iv) tax incentives for flood control, and v) types of building construction for risk areas.

The manual should also provide guidance on available potential technical alternatives for flow and water quality control, main advances and limitations. It does not need to specify how to prepare calculations for each situation, but should detail the main available methods, criteria, recommended parameters, return period and maintenance procedures, among others.

The Urban Drainage Master Plan should also include a series of specific programmes. The programmes are long-term actions which provide support to the Plan goals. The main programmes are usually: hydrologic and water quality monitoring, public participation, and capacity-building at all levels.

A hydrologic and water quality monitoring network is needed to enable estimation of model parameters used in planning, design and evaluation of control measures in urban drainage. This information reduces the cost of these actions since they are performed based on local basin behaviour.

3.6 CONCLUSIONS

Conventional approaches towards urban stormwater management tend to be based upon the rapid removal of runoff by draining floodwater in pipes and conduits. Although this may solve localized flood problems, in many situations this results in more serious flooding downstream. This is particularly noticeable in the humid tropics.

In order to compensate for this problem, urban drainage control measures should be based on appropriate management of urban impacts on the hydrologic environment. This approach is complex and includes institutional, legal and socioeconomic considerations as well as engineering aspects. However, even the more technical aspects need contemporary urban drainage designs that require much closer connection with urban planning and management of urban spaces.

It is, however, impossible to reduce urban drainage control to a simple prescription of structural and non-structural measures. The former aim to control by means of physical engineering works, and the latter by means of administrative measures that affect the physical space, and promote adaptation to living with natural hydrological phenomena. Both structural and non-structural control measures must be integrated into the environmental planning of urban areas, and are, therefore, no longer simply a problem of engineering and administrative planning.

The problems of urban development in the tropics are due not only to climatic conditions, but are also the result of density of occupation, lack of regulations and legal enforcement – mainly due to weak institutions, and a significant difference in population income, which results in illegal occupation of public or private areas.

Each problem has its own solution and each situation may require a different approach in order to minimize the impact on the living standards of the affected communities. Urban drainage and flood control plans require a good understanding of urbanization pressures and the behaviour of water in the basin.

Urban drainage control measures include a series of different aspects, ranging from technical engineering solutions, architectural design, and legal and economic questions. These aspects cannot be evaluated alone; each has to be considered from an integrated perspective, so that the adopted measures can produce the desired results.

In the humid tropics, climatic and socioeconomic conditions add particular difficulties to the use of solutions adopted in temperate areas. Problems such as greater capacity to generate runoff, greater erosive capacity, favourable conditions for the proliferation of vectors or carriers of tropical diseases, allied to uncontrolled urban expansion, precarious public works cleaning and inspection services, and technically outdated and ill-planned storm drainage systems can complicate, and even make render unfeasible, the use of certain devices and structures in use elsewhere.

In most tropical countries, the usual practice for controlling urban drainage is the rapid outflow of excess stormwater by buried conduits. The use of technical solutions to control runoff generation at its source is still, in general, at the initial stage of research and development, and is seldom applied in public or private works.

Several structural or non-structural control measures, however, can be adapted to these climatic and socioeconomic conditions if implemented within an integrated framework, in combination with environmental planning of urban areas. Although these solutions will not work correctly if not under implemented within integrated framework, it is still necessary to study isolated approaches to urban drainage control measures, applied to the humid tropics, in order to define their role in seeking technical solutions to broader environmental planning.

Contemporary drainage solutions that incorporate principles of sustainability and source control are generally seen as expensive and complex, and not suited to developing countries. This introduces inertia which constrains the advancement of alternative approaches and the improvement of urban drainage. As a result, in most developing countries, urban drainage practices do not fulfil these principles. Although, these principles have been applied in developed countries, in many cases they are also not fully implemented.

The technical tools and principles of urban drainage control which are well-known in developed countries are not always successful applied in developing countries in the humid tropics. Consequently, stormwater management measures – including both structural intervention and urban drainage policies – should be adapted according to the socioeconomic, institutional, as well as climatic and environmental, conditions prevalent in the humid tropics.

The situation is also characterized by a lack of cross-disciplinary efforts between engineers, architects and urban planners, which results in each area of specialization fearing reciprocal interference in what they see as mutually-exclusive professional domains. Therefore, the use of technical solutions aimed at reducing the impact of urbanization on flooding and other problems related to stormwater management in the tropics, remains at the initial stages of research and development.

Capacity-building should be developed at all levels: to update technical support for county/region and private engineers regarding new conceptions of flood control and urban drainage; to subside more information about source control and non-structural control measures to enable architects and engineers to design improved constructions that take these into account; and to inform the overall population in order to increase their participation in public decisions.

REFERENCES

Batista, B.B., Nascimento, N.O. and Barraud, S. 2005. Técnicas Compensatórias em Drenagem Urbana. Porto Alegre: ABRH.

Brazil, Law n.10.257, de 10 de julho de 2001. Regulamenta os Artigos 182 e 183 da Constituição Federal, estabelece diretrizes gerais da política urbana e dá outras providências, Diário Oficial da União, Poder Executivo, Brasília, DF, 17 jul. 2001.

Desbordes, M. and Servat, E. 1988. Towards a specific approach of urban hydrology in Africa. *Hydrologic Processes and Water Management in Urban Areas. Proceedings* Duisburg Conf., Urban Water 88, April: 231–37.

Goldenfum, J. A., Tassi, R., Meller, A., Allasia, D.G. and Silveira, A.L.L. 2007. Challenges for the Sustainable Urban Stormwater Management in Developing Countries: From basic education to technical and institutional issues. *Novatech 2007 – Techniques et stratégies durables pour la gestion des eaux urbaines par temps de pluie – 6ème Conférence internationale.* Vol. 1. Lyon, France: GRAIE, pp. 357–64.

Marsalek, J., Jimenez, B., Karamouz, M., Malmquist, P., Goldenfum, J.A. and Chocat, B. 2008. *Urban Water Cycle Processes and Interactions.* Vol. 1. London: Taylor & Francis.

Mecklenburg, D. 1996. *Rainwater and Land Development: Ohio's Standards for Stormwater Management Land Development and Urban Stream Protection*, 2nd edn. Columbus, Ohio, US: Ohio Department of Natural Resources.

Nicholas, D. 1995, Techniques for On-Site Stormwater Detention in Sydney, Australia: Quantity and Quality Control Mechanisms for Frequency Staged Storage, 2nd International Conference on Innovative Technologies in Urban Storm Drainage, NOVATECH 95, Lyon, France.

Ruiter, W. 1990. Watershed: Flood protection and drainage in Asian Cities. *Land & Water In'l* 68: 17–19.

Sangaré, I.B. 1998. Évaluation de quelques conditions de durabilité de l'assainissement urbain : le cas de Tours, Novatech 98, 3ème Conférence Internationale, Les Nouvelles Technologies en Assainissement Pluvial, Lyon, France, Vol. 1, pp. 173–81.

Schueler, T.R. 1987. *Controlling Urban Runoff: A Practical Manual for Planning and Designing Urban BMPs.* Department of Environmental Programs, Metropolitan Washington Council of Governments.

Silveira, A.L.L., Goldenfum, J.A. and Frendrich, R. (eds) 2001. Urban Drainage Control Measures. *Urban Drainage in Humid Tropics*, Vol. 1 . Paris: UNESCO, pp. 125–54.

Tokun, A. 1983. Current Status of Urban Hydrology in Nigeria. *Urban Hydrology. Proceedings*. Baltimore, May/June: 193–207. New York: ASCE.

Urbonas, B. and Stahre, P. 1993. *Stormwater Best Management Practices and Detention*. Englewood Cliffs, New Jersey: Prentice Hall.

WHO. 1988. *Urbanization and its implications for Child Health: Potential for Action*. Geneva: World Health Organization.

Chapter 4

Interactions between solid waste management and urban stormwater drainage

Carlos Eduardo Morelli Tucci[1,4], Jonathan Neil Parkinson[2], Joel Avruch Goldenfum[1] and Marllus Gustavo Ferreira Passos das Neves[3]

[1]Institute of Hydraulic Research, Federal University of Rio Grande do Sul, Porto Alegre, Brazil
[2]International Water Association, London, United Kingdom
[3]Centre of Technology, Federal University of Alagoas, Maceió, Brazil
[4]FEEVALE University Center, Novo Hamburgo, Brazil

4.1 INTRODUCTION

Urban areas in developing countries are frequently characterized by rapid expansion. During the process of unregulated urbanization, the lack of regulation of construction sites favours the excessive production of sediments, which are washed off with runoff. In addition, in informal settlements, unsurfaced public areas are common and these areas increase sediment concentrations in runoff.

The majority of states in the humid regions are low-income countries. These are characterized by socioeconomic and institutional factors that have significant implications on both the accumulation and control of solid wastes in urban drainage systems. The impacts of the level of economic development on the status of drainage and solid waste management services in urban areas are summarized in Table 4.1.

Many urban drainage problems are directly related to solid waste management issues and there are strong links between urban water systems and arrangements for solid waste collection and disposal. In the majority of developing countries, increasing production of refuse per capita and lack of concern about environmental issues mean that local people often use urban creeks and drainage channels as disposal sites (Silveira et al., 2001). Even where refuse is not disposed of directly into drains, solid wastes are often dumped indiscriminately and these are washed into the drainage system or are blown in by the wind.

Sam (2002) describes the problems related to ineffective municipal solid waste management in Accra, Ghana, which are typical of many developing countries. He notes that problems are encountered at all levels of waste management, namely, collection, transportation and disposal. This situation is common in many countries and, in most cases, there are not enough waste collection vehicles for effective waste management. As a result, limited existing resources are stretched to cover wider catchment areas. In addition, there is no waste separation at the source except for that undertaken by very poor scavengers, which means that hazardous and clinical wastes are often handled together with municipal solid waste.

The removal, transport and disposal of solids required to maintain the hydraulic capacity of drainage networks represents an important maintenance cost for local

Table 4.1 Urban environmental issues associated with drainage and solid waste management related to the level of economic development

	Low	Lower-middle	Upper-middle	High
Drainage	Storm drains inadequate and poorly maintained; frequent flooding, and high prevalence of water-related disease vectors.	Improved drainage in many areas but low-income areas remain unserved.	Better drainage throughout the city but occasional flooding persists. Pollution in urban watercourses remains a problem.	Good drainage results in reduced flooding. Improved control of pollution from urban runoff.
Solid waste management	Poor collection services but high demand for recycleable waste; open dumping or burning of mixed wastes; high exposure to disease vectors.	Moderate coverage of collection service, little official separation of wastes; illegal dumping and uncontrolled tipping.	Better organized collection of domestic wastes but hazardous waste management remains a problem and landfills poorly managed.	Waste reduction and resource recovery. Prevention of hazardous waste, controlled landfills or incineration.

Source: adapted from World Development Report, 2003

authorities. As solid waste management consumes the major annual budget of urban drainage operations, large-scale investments in drainage are wasted if infrastructure is not properly managed. Neglecting maintenance increases the cost of operating facilities while the deterioration of infrastructure represents an enormous loss of financial resources related to reduced asset life and premature replacement (UNCHS, 1993).

Climatic conditions in the humid tropics result in frequent flood events during the wet season due to high rainfall intensity. There are many problems that can complicate, and even make unfeasible, the use of some devices and structures used in temperate and cold climates. According to Silveira et al. (2001), these problems include:

- uncontrolled urban expansion
- a greater capacity to generate runoff
- greater erosive capacity
- precarious public works cleaning and inspection services
- technically outdated and ill-planned storm drainage systems, and
- favourable conditions for the proliferation of vectors or carriers of tropical diseases.

In the humid tropics, the volume of sediments and refuse is often so large that conventional structural solutions used by cities in developed countries are not appropriate and new approaches need to be developed (Silveira et al., 2001). Although the impacts of solids in drainage systems are well known, the ability of urban drainage engineers to develop and implement effective solutions is frequently limited by a lack of understanding of the range of different types of solid waste, and their sources, impacts and behaviour, which influence the most effective form of control strategy.

4.2 ORIGINS AND CHARACTERISTICS OF SOLIDS

The types of solid wastes generated in the urban environment which eventually find their way into drains arise from a wide variety of sources. These are summarized in Table 4.2.

There are two main stages of urban development which affect the source and nature of these solids:

1) Initial stages of development often result in a large amount of sediment because construction sites create unprotected surfaces and remove vegetal cover. The energy from rainfall intensity and high water velocity due to overland flows from impervious areas increase soil erosion and transport more sediment and vegetation into urban creeks.

2) Once the urban area has been inhabited, sediment remains the main constituent of solid waste, but other solid wastes increase due to human activity. A lack of efficiency in street cleaning and waste collection services may be compounded by a lack of awareness amongst the resident population of the need to dispose of household wastes properly.

Waste generation rates are affected by socioeconomic development, the degree of industrialization and the climate (in particular, rainfall). Generally, the greater the economic prosperity and the higher the percentage of urban population, the greater the amount of solid waste produced (see Table 4.3).

The amount of solid waste produced per person varies with population income, seasonality and regional characteristics, among other factors (Mercedes, 1997; Reis et al.,

Table 4.2 **Sources and types of solid wastes**

Source	Typical waste generators	Types of solid wastes
Residential areas	Single and multi-family dwellings.	Food wastes, paper/cardboard, plastics, textiles, leather, wood, glass, metals, ash, consumer electronics, batteries etc.
Industrial processes	Light and heavy manufacturing, construction sites, power and chemical plants, mineral extraction and processing industries.	Industrial process wastes, scrap materials, off-specification products, fabrication materials and bi-product wastes, hazardous and special wastes.
Commercial	Stores, hotels, restaurants, markets, office buildings, etc.	Paper, cardboard, plastics, wood, food wastes, glass, metals, and medical wastes.
Institutional	Schools, hospitals, prisons, government centres.	
Construction and demolition	New construction sites, road repairs, building renovation sites, demolition of buildings.	Construction and demolition materials containing: wood, steel, concrete, silt and sediment etc.
Municipal services	Street cleaning, landscaping, parks, and other recreational areas.	Street sweepings, landscape and tree trimmings, general wastes from parks and other recreational areas.

Source: Hoornweg and Thomas, 1999

Table 4.3 Solid waste production per capita per day according to level of economic development (GDP)

	kg per capita per day	
	Min.	Max.
Low-income	0.4	0.9
Middle-income	0.5	1.1
High-income	1.1	5.1

Source: Hoornweg and Thomas, 1999

2002). In the US, the mean value is $2.0\,$kg person^{-1} day^{-1} (EPA, 2004), whereas in Brazil it lies between 0.5 and 0.8, with a mean of $0.74\,$kg person^{-1} day^{-1} (IBAM, 2001). These numbers are increasing due to ongoing social and economic development, and because modern products contain more plastic.

Probably the most significant factor affecting the amount of solid waste that ends up in drains is the extent and frequency of waste collection services. In some developing countries these services do not cover the entire city because of difficulties in driving trucks along narrow streets.

Problems related to solid waste management are often found to be worst where modern technologies, such as the plastics industry, have been introduced before the development of a strong environmental lobby or a policy for the associated waste management (Armitage and Rooseboom, 2000). This is frequently the case in low and middle-income countries in the humid tropics.

Hall (1996) summarizes the factors influencing the quantity of solid waste that enters the drainage system:

- a lack of provision of bins and inadequate litter collection practices
- illicit dumping of solid wastes
- the failure by authorities to enforce penalties to deter offenders
- excessive packaging
- lack of awareness and irresponsible behaviour of residents when disposing of wastes
- low efficiency of street cleaning services
- lack of effective control of sediments on construction sites, and
- overall efficiency of refuse collection and street cleaning services.

In addition to the quantity of waste produced, the composition of the waste and sizes of particulates also vary considerably between different countries. Solid waste in low-income countries contains considerably more organic waste because of the more immediate need for recycling than in higher income countries, where the waste generally contains more paper, glass, metals and plastics (Hoornweg and Thomas, 1999). However, solid waste generally contains a much higher quantity of debris and waste from construction. Kolsky and Butler (2000) took measurements of the depth and size

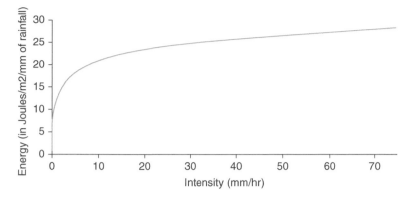

Figure 4.1 Erosive energy versus rainfall intensity

Source: Kolsky, 1998

Table 4.4 Quantities of solids generated in various urban settlements

Area description		Mass (kg ha^{-1} year^{-1})	Volume 10^{-3} (m^3 ha^{-1} year^{-1})
Springs (South Africa)	299 ha – 85% is commercial and industrial and 15% is residential.[1]	67	0.71
Johannesburg (South Africa)	8 km^2, commercial, industrial and residential.[2]	48	0.50
Sydney (Australia)	322.5 ha, commercial, industrial and residential.	22	0.23
Auckland (New Zealand)	Residential 5.2 ha Commercial 7.2 ha Industrial 5.3 ha	2.8 61.7% 26.1% 12.2%	0.03

Cape Town (South Africa)– downtown with 96% residential, 5% industrial.[3]
1. Armitage et al. (1998) 2. Allison et al. (1998) 3. Arnold and Ryan (1999)

distribution of solids in an open drain in Indore, Madhya Pradesh in India and showed that these were of a magnitude ten times greater than sewer solids in the UK.

4.2.1 Solid waste production and volumes of waste found in drains

Analyses of data collected from many storm events indicate that rainfall and runoff are the best explanatory variables for estimating gross pollutant loads (Allison et al., 1997). As shown in Figure 4.1, the higher rainfall intensities mean a greater capacity to erode and transport solids into the drainage system. Thus, the solid loading of runoff during the 'first flush' flows from urban surfaces is higher than in temperate climates. However, as shown below, there is little significant increase in the energy per mm of rainfall above an intensity of 25 mm/hr.

Neves (2006) presents a summary of the estimated mass and volume of solids generated in various urban settlements in different countries (see Table 4.4). These data present significant variability because of the various factors described above.

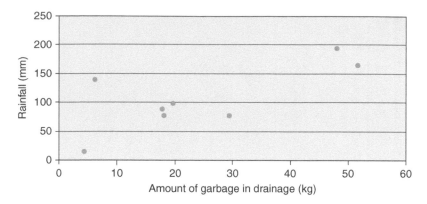

Figure 4.2 **Relationship between rainfall and amount of garbage in the drainage system (monthly values)**

Source: Neves, 2006

Neves (2006) monitored the mass balance of solid waste in a detention pond in Porto Alegre, Brazil over a period of eight months (November 2003 to June 2004). The per capita production in the basin was estimated to be $0.53\,kg\;person^{-1}\;day^{-1}$, but only a fraction of this (0.11 kg) was estimated to enter the drainage system. The amount leaving the pond was measured to be $0.034\,kg\;person^{-1}\;year^{-1}$; indicating that approximately 70% of the garbage remains in the drainage system. However, an estimated 40% was estimated to be degradable solids that are subsequently washed out of the drainage system over a period of time, leaving an estimated 30% in the drains.

Neves (2006) also estimated the amount of solids that enter the drainage system using the statistical relationship of the amount obtained on a dry day to the amount obtained on a rainy day. Figure 4.2 shows that there is an important relationship between rainfall and the amount of garbage in the drainage system.

4.2.2 Composition of solid waste in drains

The composition of refuse that arrives in the drainage system varies according to the function of the urban area, the amount of recycling and the efficiency of the services. The main types of solid waste found in urban drains are:

- domestic wastes generated by the resident population (e.g. plastic, papers etc.)
- grit, sediments and silt washed off from urban surfaces, and
- leaves and other forms of vegetation.

Table 4.5 classifies the solids found in the drains for various catchments, related to the data in Table 4.4. It can be seen that plastic is the main type of solid, largely because of its low recycling value (depending on the type of plastic), and its increased use related to the income of the population.

Using data from street cleaning activities in a catchment in Porto Alegre, Brazil, Neves (2006) estimated the volume and composition of solids entering the drainage

Table 4.5 Composition of solid waste in various cities

Local	Composition
Springs (South Africa)[1]	Plastic – 62%, Polystyrene – 11%, Paper – 10%, Cans – 10%, Glass – 2%, others – 5%.
Johannesburg (South Africa)[1]	Sediments, suspended solids (80% plastic packs), floating matter and other material.
Melbourne (Australia)[2]	80% vegetation and litter dropped by pedestrians and car drivers.
Auckland, New Zealand	Rigid plastic – 53%, Soft plastic – 1.9%, Plastic fiber – 10.5%, Glass – 0.3%, Aluminum – 3.3%, Stteel cans – 0.5%, Paper – 26.8%, other – 3.5%
Cape Town (South Africa)[3]	More than 50% plastic in industrial and commercial areas. Metal, wood and rubber also make an important contribution.

Notes 1, 2 and 3 – refer to Table 4.4

(a) Sample from street cleaning   (b) Sample from a detention pond

Figure 4.3 Samples from street cleaning and from a detention pond in Porto Alegre, Brazil

Source: Neves, 2006

system. Figure 4.3 shows a sample of the garbage in the street and another sample collected from a detention pond. The total proportion of the solids from natural contents such as sediments, wood, stones and other vegetation was 77%, compared with 23% of solids derived from domestic waste.

This last type of solid is classified in Table 4.6. The results show that plastic and paper are the main types of solid found at the drainage entrance (81%). In the basin output, there is a major change since paper-based wastes degrade and are washed out of the outlet whereas plastic remains.

In Brazil, the plastic content in waste has increased over the last few years; however, organic matter still represents between 30%–70%, paper 25%, and metal, glass and plastic account for about 10%. Plastics are generally the type of solid most prevalent in the environment. According to the US Environmental Protection Agency (EPA, 2004), plastics represent 59% of the waste in the US collected through the cleaning coast programme. Marais et al. (2004) reported that plastic is the main solid waste found in the drainage systems of cities in South Africa, Australia and New Zealand. In the study undertaken by Neves (2006), plastics with a low economic value for recycling, such as supermarket packs represent nearly 50% of the total.

Table 4.6 **Characterization of domestic solid waste found in the drainage system**

Type	Drainage entrance % of total[1]	Basin output at the detention pond % of the total
Plastics, PET and polypropylene	42	82
Paper	39	1
Shoes and tissues	3	10
Glass	5	2
Metal (aluminum and cans)	7	2
Others	4	3

Note 1: Estimated by street cleaning samples
Source: Neves, 2006

Figure 4.4 **Urban erosion**
Source: Campana, 2002

4.3 URBAN SEDIMENTS

Soil erosion in urban environments is exacerbated by the following conditions:

- high slopes and a lack of surface water drainage systems
- unprotected surfaces that increase sheetflow erosion, which in turn increase the concentration of sediment and amount of vegetation in the runoff, and
- an increase in the flow velocity from impervious surfaces, pipes and channels without hydraulic dissipation, resulting in high erosive capacity and an increase in degraded areas, as shown in Figure 4.4.

The suspended sediment yields of some urban basins in Curitiba at different stages of urbanization are presented in Figure 4.5. The Atuba Basin, which was undergoing urbanization at the time of data collection, shows the highest values of suspended solid concentration for the same specific discharge (rate of flow per area of basin). In the

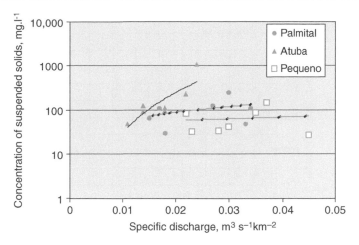

Figure 4.5 Suspended sediment concentration as a function of the specific discharge for some Curitiba basins: Atuba (15% impervious area); Palmital (7% impervious area); and Pequeno (almost rural)

Source: Tucci and Porto, 2001

Table 4.7 Volume of sediment produced in various urban basins in Brazil

River and city	Source	Volume $(m^3 km^{-2} year^{-1})$	Reference
Tietê River, São Paulo	Dredge sediment	393	Nakae and Brighetti (1993)
Tietê River tributaries, São Paulo	Bed sediment	1400	Lloret Ramos et al. (1993)
Pampulha Lake, Belo Horizonte	Sediments from lake bed	2436	Oliveira and Baptista (1997)
Dilúvio Creek, Porto Alegre	Dredge sediment	750	DEP (1993)

Palmital Basin, urban development was lower than Atuba; and the Pequeno Basin is predominantly rural. The main difference in the curves relating to the concentration of suspended solids can therefore be explained by the relationship with specific discharge, which increases in relation to the level of urbanization.

Table 4.7 shows the amount of sediment dredged from several urban rivers in Brazil.

4.4 IMPACTS OF SOLIDS ON OPERATIONAL PERFORMANCE

Sediment, sand and gravel generated by runoff from building sites and unsurfaced areas can have a significant implication on the performance of drainage systems and pollutant load discharges into receiving waters. As described below, solid wastes in drainage systems can be directly linked to:

a)   a deterioration in hydraulic performance, and
b)   an increase in pollutant loads into the environment.

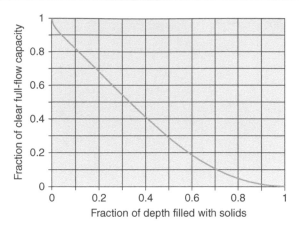

Figure 4.6 Approximate pipe capacity reductions with increasing depth of solids

Source: Kolsky, 1998

The consequences are firstly an increase in the cost of maintenance of urban drainage; secondly, environment degradation caused by erosion and transport of polluted water; and thirdly the obstruction of urban drainage channels, which causes a loss in hydraulic capacity of the drainage system and increases flood frequency.

4.4.1 Hydraulic impacts

Uncollected solid wastes enter surface drains and sewers causing blockages and reducing flow capacity. One of the main causes of flooding in urban drainage in cities in developing countries is the decreased hydraulic capacity of stormwater drainage due to solids filling conduits and channels (Tucci, 2001). As illustrated in Figure 4.6, the increasing depth of solids in the drain has an almost linear impact on the decrease in hydraulic capacity of the conduit.

Loss of hydraulic capacity can be caused by a sudden blockage from large solid wastes, or a gradual decline of hydraulic capacity caused by the deposition of smaller particulates, sediments and sands. Solids in sewers reduce the hydraulic capacity of sewers by decreasing the cross-sectional area and increasing hydraulic roughness. Solids in drains decrease hydraulic capacity, which is likely to increase the frequency and magnitude of flooding. Excessive solid deposits in the drainage system will inevitably result in operational problems, causing premature surcharge and operation of combined sewer overflows, surface flooding and overland flow of storm runoff.

Operational problems generated by a wide variety of solid wastes in urban stormwater drainage systems are widely recognized and documented (e.g. Ashley et al., 2004). The problems caused by sewer solids relate to physical effects, for example, blockages and conveyance constraints; and the effects on hydraulics related to the impact on the relative roughness of the boundary (Ashley et al., 1996). During storms, drainage structures are unable to convey stormwater due to constricted flow and reduced capacity, and consequently overflow causing floods.

Solid waste and sediment also have significant impacts on the operational performance of retention and detention tanks. Piel et al. (1999) obtained results on the performance of various storage ponds in France. More than 200 different installations were evaluated and the researchers concluded that performance deteriorates with time, particularly where ponds are covered. In the majority of situations this is caused by poor maintenance due to difficult access (blocked openings and need for professional equipment), or because installations are forgotten after multiple changes in ownership or site management. Experiences from developing countries suggest that the situation may be considerably worse in cities undergoing rapid development (e.g. Brazil – see Nascimento et al., 2000).

4.4.2 Environmental impacts of urban runoff

As well as the increase in frequency and magnitude of urban flooding, urbanization results in pollution problems in urban streams and other receiving waters. Solids are an important source of pollutant loads to river systems because organic materials and chemicals that arrive in the aquatic system are attached to solid particulates. In addition, some types of solid waste, particularly plastics, take a very long time to degrade in the environment.

Wastewater discharges during dry weather conditions are a considerable cause of pollution if untreated, but wet weather has the effect of cleaning urban surfaces and drainage channels, resulting in significant pollution problems. The quality of runoff is influenced by many factors such as land use, waste disposal and sanitation practices. In the humid tropics, pollution problems are worse due to the fact that higher temperatures dissolve more solids and reduce the oxygen concentration in the receiving waters.

A significant amount of pollutants, ranging from gross pollutants to particulates and soluble toxins, are generated from urban catchments. There are a number of pollutants of principal concern in urban runoff, and these affect organisms in receiving waters in various ways (see Table 4.8). It is important to note that these environmental stressors may interact to varying degrees in antagonistic, additive or synergistic ways (Porto, 2001). This means that the cumulative effect of different pollutants is likely to be worse than the sum of the individual effects of each one.

As described above, high rainfall intensities have a particularly high erosion capacity. Consequently, suspended solids concentrations in runoff can be very high, which can cause significant pollution problems. During rainstorms, deposited solids – particularly sediments and small organic particulates – are resuspended and discharged into receiving waters.

The deposition of sewage solids during dry weather in combined sewers has long been recognized as a major contributor to the 'first-flush' phenomena. This involves the mobilization of loose solids on the urban ground surface that are transported into the sewerage system by surface storm runoff, and the scouring of materials already deposited in the drainage system and ancillary structures such as retention tanks.

These particulates may settle out in the system and be resuspended during wet weather periods, which generates a 'first-flush' loading from storm drainage systems. Ashley et al. (1996) concluded that the quality and potential pollution problems of first flushes linked to erosion or mobilization of the deposited solids and consequences for CSO discharges and shock loads on treatment plants are significant, while control strategies are still poorly developed.

Table 4.8 Environmental impacts of solid waste pollutants in urban stormwater runoff

Pollutant	Source	Environmental impact
Oxygen-demanding materials	Vegetation, excreta and other organic matter	Depletion of dissolved oxygen concentration kills aquatic flora and fauna (fish and macro-invertebrates) and caused imbalance in the composition species in the aquatic system. Odours and toxic gases form in anaerobic conditions.
Inorganic compounds of nitrogen and phosphorus	Fertilizers, detergents, vegetation, animal and human urine, sewer overflows and leaks, septic tank discharges.	In high concentrations, ammonia and nitrate are toxic. Nitrification of ammonia microorganisms consumes dissolved oxygen. Nutrient enrichment (eutrophication) causes excessive weed and algae growth blocks sunlight, which affects photosynthesis and causes oxygen depletion.
Suspended solids, sediments and dissolved solids.	Erosion from construction sites, exposed soils, street runoff and stream banks.	Sediment particles transport other pollutants attached to their surfaces. Sediments interfere with photosynthesis, respiration, growth and reproduction and deposited sediments reduce the transfer of oxygen into underlying surfaces.
Refuse and debris	Domestic and commercial refuse, construction waste and various types of vegetation.	Blockages and constrictions to drainage channels, aesthetic loss and reduction in recreational value.

Source: Parkinson and Mark, 2005

Runoff from roads and other paved areas is of concern as it harbours a vast array of particulates and chemicals arising from activities that characterize land use. Roads and other transport-related impervious surfaces are of particular concern as they contribute a higher proportion of stormwater pollutants than other impervious surfaces (e.g. roof areas and pedestrian pathways). Runoff from transport-related surfaces consistently shows elevated concentrations of suspended solids and associated contaminants such as lead, zinc and copper, as well as other pollutants such as hydrocarbons (Wong et al., 2000).

In addition, problems related to microbiological pollution are caused by the flooding of sanitation systems and the discharge of faecal solids containing pathogenic bacteria (faecal coliforms, faecal streptococci and enterococci) and other microorganisms (viruses, protozoa, etc.) that can cause intestinal infections.

4.5 SOLIDS MANAGEMENT

Managing solids in the urban environment requires actions into two areas:

- management of total solids yields, including sediment yield and domestic solid waste production, and
- control of solids that enter the drainage system, by use of engineering solutions, drain cleaning or improved solid waste collection and disposal.

Total solids in an urban environment is given by the sum of the following:

$$TR = Tc + Tl + Tdr$$

where
TR is the total solids in weight (kg) produced in a period of time
Tc is the amount collected in buildings
Tl is the total obtained by the street cleaning, and
Tdr is the amount arriving in the drainage system.

The value of TR increases with the population income. This means that the management of the first two terms on the right-hand side of the equation must become more efficient, in order to have a minimum value of Tdr. A high value of Tdr results in higher cleaning costs, greater impact on the environment and an increase in the frequency of floods.

The amount of the two first terms on the right-hand side of the equation requires a disposal site with a controlled environmental impact. However, it could be decreased by recycling, thus reducing the amount that needs to be disposed. When these services are not efficient, the third term on the right-hand side of the equation increases, causing impacts on the environment (Tdr).

In addition, when considering strategies to overcome the problems described above, it is important to consider the different types and constituents of solid waste.

4.5.1 Sediments yield

Management of sediment yields in an urban environment requires the development of regulation for new constructions in urbanization areas. Regulation in some developed countries, such as the US, is based on the use of detention ponds and other small hydraulic structures in creeks downstream of the urban surface in order to dissipate the flow energy.

4.5.2 Improved solid waste collection and disposal

Although some form of drain cleaning will always be necessary, it is not always the most efficient or cost-effective approach and should not be the only strategy adopted for the management of solids.

The most effective strategy for reducing solid waste is to stop it entering the drainage system in the first place, and the most effective way to do this is to ensure that there is an efficient solid waste collection system operating in the city. In this way, street and drainage cleaning are closely related.

The sweeping of urban surfaces can also be important for the control of solid waste entering the drainage system. However, its effectiveness as a sediment control strategy is limited. Increasing the frequency of street sweeping practices beyond what is required to meet aesthetic objectives is not expected to yield substantial incremental benefits in relation to improvements in the receiving water quality. This is because many of the pollutants are attached to finer sediments, which are not collected by most sweeping practices (Walker and Wong, 1999). In San Jose, California, US, the amount of refuse in the storm network was reduced to $0.8\,kg\ person^{-1}\ year^{-1}$ after street cleaning was improved (Larger et al., 1977).

Box 4.1 Exchange programme

In Curitiba, Brazil, a programme was developed for low-income neighbourhoods to exchange solids collected by the population for bus tickets, which created an economic value for the solids collected by the population. The cost of the bus ticket promotion for the municipality was lower than the cost of the maintenance. The bus ticket did not increase the cost, since the city pays for the number of buses in the streets and not for each ticket.

Recycling is a function of the population's awareness in selecting its waste before it is collected. In developing countries, the main incentive is economic, since part of the income of the poor population comes from recycling items such as aluminum cans and other profitable metals, paper and plastic. Box 4.1 presents an example of this economic incentive.

4.5.3 Engineering solutions for control of solid wastes in drains

Drainage of urban runoff requires the construction of drains in the form of open channels or pipes buried in the ground. However, there are alternative approaches based on the use of overland flow channels to drain large-scale floods. 'Drainage without drains', a phrase coined by Kolsky et al. (1996), is an important concept for the management of urban stormwater runoff, particularly in the humid tropics where high intensity rainfall often causes flash flooding.

In the urban environment, this essentially involves the use of roads as conduits for runoff during large storm events. This approach means that the designed discharge capacity of drains themselves can be reduced and the drains have a lower volumetric capacity for the accumulation of solid wastes. In addition to this, the problem becomes more apparent to local residents if solid waste is not removed, and it is easier to remove solid waste from roads.

As described below, there are a number of design features that should be included in drainage systems to control solids deposition.

Inlet solid traps and gulley pots

One of the most important design features of closed drainage systems is the type of inlet. In closed drainage systems, inlets are where runoff enters the system. However, depending on the design of the inlets, they can also stop the ingress of larger solids into the drainage system through the inclusion of sediment traps and smaller apertures. This is desirable, but the greater the efficiency of the inlets with regard to stopping solids entering the drains, the higher the chances of inflow of runoff being constricted – especially if the inlets become blocked, either partially or fully. As a consequence, flooding is more likely to occur.

Therefore, for covered or closed systems to be efficient at draining stormwater runoff, they need to have inlets that are large and numerous enough to allow for the inflow but also effective at stopping the ingress of solid waste.

In addition, they must not be so big that they become dangerous to local residents (particularly children). Various types of inlet control and sediment or solids traps are available, which contribute towards reducing the ingress of solids into the drainage system. Inlets may include a wide range of devices such as grids or gully pots. However, if they are effective at keeping solids from entering the drainage system, the inlets invariably become blocked.

It is of vital importance that there is a routine maintenance system to ensure that the inlets are regularly cleaned – particularly before and during the rainy season. Butler and Clark (1995) highlight the importance of the gully pot as an effective sediment trapping mechanism. If a routine cleaning operation is implemented, these devices have the greatest potential for overall cost-effectiveness for a sediment control strategy in a catchment area.

4.5.4 Drain cleaning

Although the technical recommendations and engineering solutions described above can be effective in reducing solids loads into the drainage system, it is not feasible for a drainage system to remain free from problems related to blockages or loss of capacity due to deposition of solids. Therefore, drain cleaning is an important component of any strategy for reducing problems relating to solids in drainage.

The effectiveness of drain-cleaning operations depends on ease of access to the drains with suitable equipment; this in turn depends on the design of the drains. Closed drains are much harder to clean, unless they have been designed to allow the easy removal of covers; the distance between access chambers is also important.

Drain cleaning accounts for by far the most common management problem related to solid waste in drains, particularly in developing countries. But in the majority of situations, drain cleaning is infrequent and usually takes place only in response to blockages. Drain cleaning either requires labour or specialist equipment and a number of conventional cleaning techniques are described below, followed by a discussion of various manual and automated flushing methods.

Access for cleaning

The more inaccessible the location, the higher the cost of cleaning. Thus, surface drains are generally easier and cheaper to construct compared to buried pipe systems. For a closed system, the distance between manholes and access chambers has a significant influence on the ease of access for cleaning. Another important aspect related to access is the location of the drains in relation to the road network – drains laid where access by road is difficult are consequently more difficult to access with drain-cleaning equipment.

In-system solids collection

Solid waste traps may be installed at strategic locations in stormwater drainage systems to collect and remove solid waste from the flow. There are many different types of solid waste trap, and their efficiency at removing solids is highly variable. The majority rely on screening the stormwater runoff, vortex flow to remove solids by centrifugal force, or surface debris traps to remove floating debris. The main criteria for determining the suitability of a particular trap are the flow rate, allowable head loss, size, efficiency, reliability, ease of maintenance and cost effectiveness. However, the choice of trapping structure is site-specific and the location of the traps is crucial.

Decisions need to be made to include one or two large solid traps towards the end of the drainage system or to install smaller ones at strategic points in the network. Efficiency will decrease rapidly if these traps are not properly cleaned and maintained, therefore easy access for cleaning and maintenance is crucial. In some instances, the cost of providing adequate access may be higher than the structure itself (Armitage and Rooseboom, 2000).

4.6 INSTITUTIONAL FRAMEWORK FOR SOLID WASTE MANAGEMENT

Many infrastructure improvement projects direct considerable attention to design and construction, but few pay sufficient attention to long-term operation and maintenance requirements. Deficiencies in operation and maintenance inevitably mean that drainage systems do not operate in the way in which they were designed, and consequently the expected level of benefit is not achieved (Parkinson and Mark, 2005). It is therefore recommended that any infrastructure improvement project should incorporate the design of the maintenance strategy and, where necessary, this design should be adapted to take into account specific maintenance procedures.

The different types of equipment available for the cleaning of drainage systems are rarely considered in the design of drainage systems, which is based on the assumption that drainage systems operate under solid-free conditions. It is evident that this is rarely, if ever the case. It is also particularly important to prioritize drain cleaning programmes. This may involve the evaluation of system performance monitoring to identify which parts of the drainage system are causing the most significant problems (Kolsky and Butler, 2002).

Experience of drainage system operation suggests that certain sections of drainage systems are regularly acting as points for solids deposition; it might therefore be beneficial to focus the drain cleaning on these points on a regular basis as oppose to attempting to clean the whole of the drainage system. However, further research is required to provide guidance for design engineers on the effectiveness of different maintenance strategies. There is also a related need for field-based research to characterize and identify the most effective equipment for removing different types of solid waste from drainage systems, taking into account the differing composition of solid waste in drains in developing countries.

There is a need for better understanding of the processes related to solid wastes in urban stormwater drainage systems in the humid tropics. The solutions developed under other conditions can be adapted to climatic and socioeconomic conditions within an integrated framework, together with the environmental planning of urban areas, combining all aspects involved. It is particularly important to develop solid waste management strategies that are appropriate for the local socioeconomic and institutional context. For instance, it might be appropriate to seek ways in which the municipality can develop partnership management arrangements with community groups, NGOs or the private sector. These partnerships, although potentially more complex to set up, can prove to be more effective in service delivery than single organizations attempting to do everything.

It is logical that the organization responsible for solid waste management should also be responsible for drainage, as operation of the drainage system is intrinsically linked to control of solids wastes. Moreover, it is more effective and easier to stop waste entering the drainage system.

The local authority is probably the most appropriate agency to take on this role. But there are considerable advantages in developing partnerships with the private sector when the key success factors of competition, transparency and accountability are present. Lease contracts offer clear incentives and opportunities to reduce costs by introducing competition, provided the cost of cleaning is not set by local government. The private sector improves efficiency and lowers costs by introducing commercial principles such as limited and well-focused performance objectives, financial and managerial autonomy, and clear accountability to both customers and providers of capital. The private sector may also play an important role in providing new ideas, technologies and skills, which also mean that the need for the municipality to employ a large number of staff for these activities is reduced.

One form of private sector involvement that creates a livelihood for the urban poor is the incorporation of micro-enterprises and informal waste recycling cooperatives into the municipal solid waste management system. Research has shown that the promotion of micro-enterprises is an effective way of extending affordable services to poor urban communities. The promotion and development of recycling cooperatives also provides a way of upgrading the living and working conditions of informal waste pickers, resulting in higher incomes and greater self-esteem (Brook and Irwin, 2003).

REFERENCES

Allison, R., Chiew, F. and McMahon, T. 1997. *Stormwater Gross Pollutants*. Cooperative Research Centre for Catchment Hydrology. Industry Report 97/11. Monash University, Australia.

Allison, R.A., Walker, T.A., Chiew, F.H.S., O'Neill, I.C. and McMahon, T.A. 1998. *From roads to rivers – gross pollutant removal from urban waterways*. Research Report for the Cooperative Research Centre for Catchment Hydrology, Australia.

Armitage, N.P. and Rooseboom, A. 2000. The removal of urban litter from stormwater conduits and streams. Paper 1 – the quantities involved and catchment litter management options. *Water SA*, Vol. 26. No. 2, pp. 181–87.

Armitage, N., Rooseboom, A., Nel, C. and Townshend, P. 1998. *The removal of urban litter from stormwater conduits and streams*. Water Research Commission. Report No. TT 95/98, Pretoria.

Arnold, G. and Ryan, P. 1999. *Marine Litter Originating from Cape Town's Residential, Commercial and Industrial Areas: The connection between street litter and stormwater debris. A co-operative community approach*. Island Care New Zealand Trust, C/- Department of Geography, The University of Auckland, New Zealand. Percy FitzPatrick Institute, University of Cape Town.

Ashley, R., Verbanck, M., Bertrand-Krajewski, J-L., Hvitved-Jacobsen, T., Nalluri, C., Perrusquia, G., Pitt, R., Ristenpart, E. and Saul, A. 1996. Solids in sewers – the state of the art. *7th International Conference on Urban Storm Drainage*, Hannover, Germany.

Ashley, R.M., Bertrand-Krajewski, J.L., Hvitved-Jacobsen, T. and Verbanck, M. (eds) 2004. *Solids in Sewers – Characteristics, Effects and Control of Sewer Solids and Associated Pollutants*. Scientific & Technical Report No. 14. London: IWA Publishing.

Brook, P.J. and Irwin, T.C. (eds) 2003. *Infrastructure for Poor People: Public Policy for Private Provision*. Washington DC: World Bank.

Butler, D. and Clark, P. 1995. *Sediment Management in Urban Drainage Catchments*. CIRIA Report No. 134. London, Construction Industry Research Information Association.

Campana, N. 2002. *Urban Drainage*. Presentation at a meeting at the University of Brasilia.

EPA. 2004. *A Pollution Prevention Management Measure. Polluted Runoff (Nonpoint Source Pollution)*. US Environmental Protection Agency. http://www.epa.gov/OWOW/NPS/MMGI/Chapter4/ch4-6.html.

Hall, M. 1996. *Litter Traps in the Stormwater Drainage System*. Unpublished M.Eng. paper. Swinburne Univ. of Technol., Melbourne (cited by Armitage and Rooseboom, 2000).

Hoornweg, D. and Thomas, L. 1999. *What a Waste: Solid Waste Management in Asia*. Urban and Local Government Working Paper Series No. 1. Washington DC: World Bank.

IBAM. 2001. Gestão integrada de resíduos sólidos: *Manual de gerenciamento integrado*. Rio de Janeiro.

Kolsky, P.J. 1998. *Storm Drainage: An Engineering Guide to the Low-cost Evaluation of System Performance*. London, Intermediate Technology Publications.

Kolsky, P.J. and Butler, D. 2000. Solids size distribution and transport capacity in an Indian drain. *Urban Water*, Vol. 2, No. 4, pp. 357–62.

Kolsky, P.J. and Butler, D. 2002 Performance indicators for urban storm drainage in developing countries. *Urban Water*, Vol. 4, No. 2, pp. 137–44.

Kolsky, P.J., Parkinson, J.N., Butler, D. and Sihorwala, T.A. 1996. *Drainage without Drains?* Performance Studies in India and their Implications, Proceedings of the Seventh International Conference on Urban Storm Drainage, pp. 521–26. SuG-Verlagsgesellschaft: Hannover.

Larger, J.A, Smith, W.G., Lynard, W.G., Finn, R.M. and Finnemore, E.J. 1977. *Urban Stormwater Management and Technology: update and user's guide*. US EPA Report – 600/8-77-014 NTIS N. PB 275654.

Marais, M., Armitage, N. and Wise, C. 2004. The measurement and reduction of urban litter entering stormwater drainage systems: Paper 1 – Quantifying the problem using the city of Cape Town as case study. *Water SA*. No. 4. Vol. 30. http://www.wrc.org.za (accessed 5 September 2006).

Mercedes, S.S.P. 1997. Perfil da geração de resíduos sólidos domiciliares no município de Belo Horizonte no ano de 1995. 19° Congresso Brasileiro de Engenharia Sanitária e Ambiental. Foz do Iguaçu: Abes.

Nascimento, N.O., Ellis, J.B., Baptista, M.B. and Deutsch, J-C. 2000. Using detention basins: operational experience and lessons. *Urban Water*, Vol. 1, No. 2, pp. 113–24.

Neves, M. G. F. P. 2006. Quantificação de resíduos sólidos na drenagem urbana. Tese de doutorado. Universidade Federal do Rio Grande do Sul, Porto Alegre, Brazil.

Parkinson, J. and Mark, O. 2005. *Urban Stormwater Management in Developing Countries*. IWA Publishers.

Piel, C., Perez, I. and Maytraud, T. 1999. Three examples of temporary stormwater catchments in dense urban areas: a sustainable development approach. *Wat. Sci.Tech.*, Vol. 39, No. 2, pp. 25–32.

Porto, M.F.A. 2001. Water quality of overland flow in urban areas. C.E.M. Tucci, *Urban Drainage in Specific Climates: Urban Drainage in Humid Tropics*. Paris: UNESCO. IHP-V. Technical Documents in Hydrology. No 40. v.I. cap.4, pp. 103–24.

Reis, M.F.P., Ellwanger, R.M., Pescador, F.S., Cotrim, S.L., Reichert, G.A.E and Onofrio, E.T. 2002. Estudos preliminares para caracterização dos resíduos sólidos domiciliares do município de Porto Alegre. *VI Seminário Nacional de Resíduos Sólidos: R.S.U. especiais*. Gramado: ABES.

Sam, P.A. Jr. 2002. Are the municipal solid waste management practices causing flooding during the rainy season in Accra, Ghana, West Africa. *African Journal of Environmental Assessment and Management*, Vol. 4, No. 2, December 2002, pp. 56–62.

Silveira, A.L.L, Goldenfum, J.A and Frendrich, R. 2001. Urban drainage control measures. C.E.M. Tucci, *Urban Drainage in Specific Climates: Urban Drainage in Humid Tropics*. IHP-V. Technical Documents in Hydrology. No. 40, Vol. 1. Paris: UNESCO, pp. 125–56.

Tucci, C.E.M. 2001. Urban drainage issues in developing countries. C.E.M. Tucci, *Urban Drainage in Specific Climates: Urban Drainage in Humid Tropics*. IHP-V. Technical Documents in Hydrology. No. 40, Vol. 1. Paris, UNESCO, pp. 23–40.

Tucci, C.E.M. and Porto, R.L. 2001. Storm hydrology and urban drainage. C. Tucci, *Humid Tropics Urban Drainage*. Chapter 4. UNESCO.

UNCHS. 1993. *Maintenance of Infrastructure and its Financing and Cost Recovery*. Nairobi, Kenya: United Nations Centre for Human Settlements.

Walker, T.A. and Wong, T.H.F. 1999. *Effectiveness of Street Sweeping for Stormwater Pollution Control*. Cooperative Research Centre for Catchment Hydrology. Technical Report 99/8, Australia.

Wong, T., Breen, P. and Lloyd, S. 2000. *Water sensitive road design options for improving stormwater quality of road runoff*. Technical Report for Cooperative Research Centre for Catchment Hydrology 00/1.

World Bank. 2003. *World Development Report 2003. Sustainable Development in a Dynamic World*. Washington DC: World Bank.

Chapter 5

Control of public health hazards in the humid tropics

Jonathan Neil Parkinson[1] and Luiza Cintra Campos[2]

[1]International Water Association, London, United Kingdom
[2]Department of Civil, Environmental and Geomatic Engineering, University College London, United Kingdom

5.1 INTRODUCTION

Water is a mechanism for the transmission of many pathogens and other microbial and chemical pollutants. A wide variety of these pollutants contaminate water sources, affecting the potability of the water and potentially causing a range of illnesses and diseases. This chapter concentrates on pollutants that adversely affect human health, and discusses ways to mitigate potential public health risks associated with urban water systems in the humid tropics.

Water and sanitation-related diseases are one of the main causes of morbidity and mortality throughout the world. According to the World Health Organization (WHO), the number of types of diseases is expanding and the incidence of many microbial diseases is also increasing. The incidence is higher in cities located in the humid tropics and in changing environments that are linked to intensified water resources development. In addition, urbanization and the accompanying demographic changes create conditions where vector-borne diseases can gain new strongholds (WHO, 2003a).

Many urban dwellers in the humid tropics regularly suffer from various types of water and sanitation-related diseases, many of which cause endemic diarrhoea. These are most apparent in poor communities, due to inadequate water supplies, sanitation, drainage and waste collection systems. It is in situations such as these that microbial pathogens and disease vectors thrive.

Two predominant climatic factors influence the high prevalence of diseases in the humid tropics:

- Temperature – Enteric microorganisms that inhabit the human gut are *mesophilic*, meaning that they have an optimum growth temperature between 30°C and 40°C. In colder climates, pathogens die off relatively quickly in the environment once they have been excreted from the human body. However, the warmer ambient temperatures in the humid tropics are conducive to their survival and, in some instances, to the regrowth of microorganisms.
- Rainfall and humidity – Soil moisture content and humidity are important factors that influence pathogen survival in the environment. It is common to observe

increasing incidence of disease during the wet season. Higher occurrences of diseases are also intrinsically linked to deterioration in environmental health conditions caused by flooding of latrines and overflowing drainage channels, which are common during the wet season.

5.2 HEALTH RISKS ASSOCIATED WITH MICROBIOLOGICAL AND CHEMICAL POLLUTANTS

5.2.1 Infectious microbial pathogens

Microbial pathogens lead to a range of potential health hazards (summarized in Table 5.1), which are described in more detail below.

Bacterial infections – The most common bacterial infections in the faecal–oral transmission route are caused by pathogenic *E. coli.*, which cause diarrhoea and other gastro-intestinal illnesses. The bacteria *Vibrio cholerae* and *Salmonella Typhi* also transmit diseases – namely, cholera and typhoid, respectively. Other bacterial diseases include Legionnaires' disease, which is caused by the *Legionella* bacteria and causes pneumonia (a type of lung infection).

Protozoan infections – *Giardia* and *Cryptosporidium* protozoa are major causes of human morbidity and are a significant contribution to mortality (Horan, 2003). Both

Table 5.1 **Water-related pathogens and their health impacts**

Description	Name of pathogen	Health impact
Pathogenic bacteria	*Escherichia coli, Vibrio cholerae, Salmonella, Campylobacter, Shigella, Helicobacter pylori, Legionella*	Short-term effects such as diarrhoea, cramps, nausea, headaches. *Chlorella* is most severe and may be fatal *Legionella* causes pneumonia.
Protozoan parasites	*Giardia lamblia*	Gastro-intestinal illness with symptoms of diarrhoea, vomiting and abdominal cramps.
	Cryptosporidium parvum	A mild gastro-intestinal disease, but can be fatal for people with weakened immune systems.
	Entamoeba histolytica	Amoebic dysentery and liver abscesses.
	P. falciparum (and other *Plasmodium* protozoa)	Malaria – lassitude, headache, nausea and vomiting and fever.
Viruses	Rotaviruses, Caliciviruses, Astroviruses	Diarrhoea and vomiting.
	Hepatitis viruses	Tiredness, fever and diarrhoea.
	Enteroviruses	Extra-intestinal symptoms such as meningitis, headache and paralytic poliomyelitis.
	Flavivirus	Dengue and dengue hemorrhagic fever.
Intestinal parasitic nematode worms	Roundworm (*Ascaris lumbricoides*)	Intestine blockage resulting in severe abdominal pain and vomiting.
	Whipworm (*Trichuris trichiura*)	Intermittent stomach pain, bloody stools, diarrhoea and weight loss.
	Hookworm (*Necator americanus* and *Ancylostoma duodenale*)	Diarrhoea, vague abdominal pain, intestinal cramps, colic and nausea.

Source: information from various chapters in Mara and Horan, 2003

are protozoan parasites transmitted by the faecal–oral route and affect the gastro-intestinal tract of humans. *Cryptosporidium* oocysts have a hard shell, which makes them highly resistant to conventional disinfection methods such as chlorination. *Entamoeba histolytica* is another intestinal protozoan parasite that can lyse cells and destroy human tissue (Stanley, 2001).

Viruses – Viruses are a common cause of gastro-intestinal infection. The main sources of contamination of drinking water are faeces and untreated wastewater, which often contain high concentrations of pathogens. Viruses are minute and can pass through most soils. They are also highly persistent in the environment and can therefore penetrate into groundwater.

Parasitic worms (Helminths) – Faecally contaminated wet soils are ideal living conditions for intestinal parasitic nematode worms such as roundworm (*Ascaris lumbricoides*), whipworm (*Trichuris trichiura*) and hookworm (*Ancylostoma duodenale* or *Necator americanus*). *Ascaris* and *Trichuris* infections are acquired by the ingestion of the embryonated eggs (nematodes), whereas hookworm enters the body through the skin.

5.2.2 Diseases transmitted by mosquitoes

Some of the most important environmental health issues in cities in the humid tropics relate to diseases spread by mosquitoes. Both undrained runoff and stagnant waters provide breeding habitats for different types of mosquitoes.

Malaria – Malaria is the most prolific mosquito-borne disease. It is caused by the plasmodium protozoa, which is transmitted to humans by *Anopheles* mosquitoes. Although generally associated with rural areas, urban malaria is an increasing problem in peri-urban areas and is a major problem in South Asia and Africa (Lines, 2002). In the past, the use of insecticides largely controlled the transmission of malaria, but with increased vector resistance to insecticides, the disease has reappeared (Beumer et al., 2002).

Dengue fever – *Aedes aegypti*, which is the main vector for dengue, is a domestic, day-biting mosquito that feeds predominantly on human blood. Dengue (and dengue hemorrhagic) fever is one of the most widespread mosquito-borne viral diseases affecting humans, and its global distribution is comparable to that of malaria (CDC, 2005). An estimated 2.5 billion people live in areas at risk from epidemic transmission and the majority of these live in urban areas of Latin America, South and South-East Asia (Lines, 2002).

Yellow fever – The *Aedes aegypti* mosquito is also the vector that transmits the virus that causes urban yellow fever. The virus is passed directly from one human to another by a biting mosquito (horizontal transmission); or the virus is passed, via infected eggs, to its offspring, which then bite humans (vertical transmission). The mosquito eggs are resistant to drying and lie dormant during dry conditions. When the rainy season begins, the eggs hatch, ensuring that transmission continues from one year to the next.

Filariasis – Filariasis is a parasitic worm infection that causes the disfiguring and disabling disease elephantiasis, which is prevalent in many African towns and cities. It is transmitted by a variety of mosquito species, but most commonly by *Culex quinquefasciatus*. It is well adapted to urban conditions; largely because the polluted water in which it breeds is abundant in urban areas, for example, pit latrines, soakage pits, septic tanks and blocked drains (Lines, 2002).

5.2.3 Diseases transmitted by animals

Leptospirosis – Leptospirosis, otherwise known as Weil's disease, is a bacterial disease transmitted to humans by the ingestion of water contaminated by urine from infected rats or other mammals. Vulnerability to the disease is increased by inadequate garbage disposal, improper housing in flood-prone areas and poor drainage of rainwater (CDC, 2007).

Schistosomiasis – Schistosomiasis is caused by parasitic flatworms known as flukes, which have external suckers for attaching to an animal host. Their complex life cycle involves fresh-water snails that breed in standing water as intermediate hosts. The snails release large numbers of minute, free-swimming larvae (cercariae), which penetrate the skin of humans in contact with the water. Even brief exposure to contaminated water (e.g. during wading or bathing) may result in infection. Outbreaks of schistosomiasis often occur during floods.

5.2.4 Chemicals and micropollutants

Nutrients – Nitrogen pollution of groundwater is increasingly becoming a problem in developing countries (Fields, 2004). The main source is runoff from agriculture, but nitrogen in domestic wastewater is another major contributor. Nitrate levels have risen in some countries to the point where more than 10% of the population is exposed to concentrations in drinking water that are above the WHO guideline of 10 mg/l (WHO, 2003b). This guideline value aims to prevent methaemoglobinaemia, to which infants are particularly susceptible.

Cyanobacteria (blue-green algae) – The presence of cyanobacterial algal blooms that occur in stagnant reservoirs and lakes used for drinking water abstraction or for recreational purposes present an increasing health risk in the humid tropics (Kuiper-Goodman et al., 1999). These risks are particularly common in surface water bodies, which become eutrophied due to the build up of nutrients.

Cyanobacteria produce a variety of toxins, but cyclic peptides (such as microcystins and nodularins) are the most widespread and are particularly important with respect to drinking water treatment (Sivonen and Jones, 1999). Acute intoxication by cyanotoxins can result in sudden and severe liver damage (Runnegar and Falconer, 1982) and have been shown to promote tumours (Fujiki and Suganuma, 1999).

Endocrine disrupting chemicals – Some endocrine disrupting chemicals (EDCs) such as oestrodiol and oestrogen are naturally produced by humans. Additionally, some man-made substances (pesticides and organohalogens such as PCBs and dioxins) can mimic or interfere with natural hormones in the body and may result in various problems associated with reproductive and sexual development and function. EDCs may also affect the development of the nervous and immune systems and some evidence suggests that exposure to EDCs may cause cancer (US-EPA, 2001).

Pesticides – Pesticides applied to agricultural land can contaminate groundwater or surface waters used for the abstraction of drinking water supplies. Depending upon the type of pesticide and quantity consumed, pesticides can cause a wide range of health problems such as birth defects, nerve damage and cancer (US-EPA, 1999).

Heavy metals – Heavy metals are natural elements that cannot be degraded or destroyed. The heavy metals that are linked most often to human poisoning are arsenic, cadmium, lead and mercury (Table 5.2). Most heavy metals enter the water

Table 5.2 Heavy metals – sources and impacts

Heavy metal	Source	Health impact
Arsenic	Arsenic is a naturally occurring element in ground rock and soil and can be present in groundwater. Arsenic contamination of drinking water has been reported in many countries (notably Bangladesh).	Long-term exposure to arsenic via drinking water causes various types of cancer and often pigmentation changes and thickening of the skin. (*hyperkeratosis*) (WHO, 2001).
Cadmium	Cadmium may be present as an impurity in the zinc of galvanized pipes or cadmium containing solders in fittings, water heaters, water coolers and taps.	Cadmium accumulates in the kidneys and is linked to a range of kidney diseases (WHO, 1997).
Lead	Lead has traditionally been used for household plumbing and water distribution systems. Other sources include unleaded petrol, batteries and a variety of industrial contaminants.	Lead has numerous acute and chronic adverse effects (e.g. neurological damage), particularly in infants and children (WHO, 2003c).
Mercury	The use of mercury in industrial processes is decreasing because of environmental concerns and environmental legislation in many countries.	The principal health risks associated with mercury are damage to the nervous system and deformities in children exposed in the womb (WHO, 1997).

supply via the pollution of water bodies from industrial waste, but acidic rain can also break down some soils to release naturally occurring heavy metals (Elankumaran et al., 2003).

5.3 TRANSMISSION PATHWAYS IN URBAN WATER SYSTEMS

In order to derive effective strategies to mitigate health risks it is important to better understand the transmission pathways of pathogens and micro-pollutants in relation to the different components of the urban water system. These are illustrated in Figure 5.1 and described in the following section.

5.3.1 Contamination of water sources

Water pollution arises from *point sources* (e.g. effluent discharges) or from *non-point-sources*, including on-site sanitation systems and runoff from urban and agricultural land. Polluted surface watercourses may be used for a variety of purposes, such as a source of drinking water, for the abstraction of water for irrigation, for bathing or for washing. These activities present different levels of health risk.

Groundwater pollution generally arises from non-point sources, including on-site sanitation (pit latrines, cesspools, septic tanks with soakaways) as well as leakages from joints or pipe fractures in sewer pipes into the surrounding soils (see Figure 5.2).

Determining the movement of microorganisms in soil is complex, but viruses penetrate deeper than larger microbes. Therefore, the clay content of the unsaturated zone is a critical indicator of the likely mobility of contaminants and their subsequent impact on groundwater pollution (Cotton and Saywell, 1998).

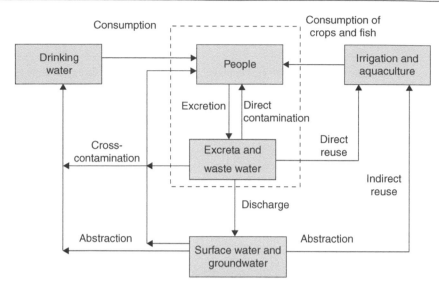

Figure 5.1 **Main pathways of human exposure to pathogens in the aquatic environment**

Source: adapted from Bosch et al., 2001

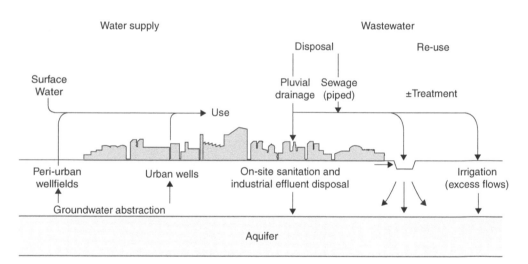

Figure 5.2 **Interaction of groundwater supply and wastewater disposal in a city overlying a shallow aquifer**

Source: Foster et al., 1998

Groundwater pollution is of particular concern from a public health perspective where groundwater is abstracted for water consumption in the following situations:

- *Contamination of shallow surface groundwater* – Although the risks may sometimes be overly dramatized compared with the risks of disease transmission by other routes, contaminated groundwater is frequently a source of water-borne

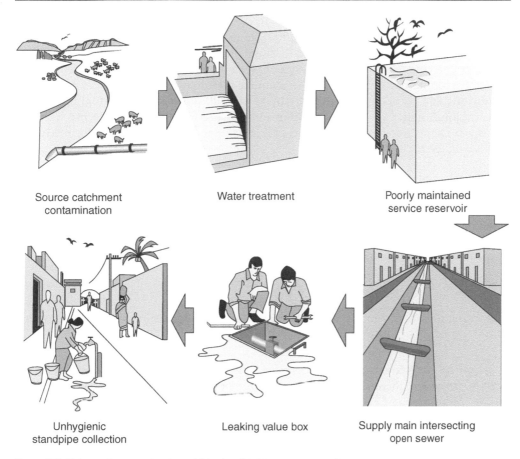

Figure 5.3 **Points of contamination within the drinking water supply system**

Source: Godfrey and Howard, 2004

transmission where residents abstract water from shallow wells. This is particularly common in cities where water supplies are unreliable and intermittent and low-income residents cannot afford to purchase water from vendors.

- *Contamination of deep aquifers* from mobile pollutants (e.g. nutrients and micro-pollutants) arises from the infiltration of polluted shallow water. Contamination is often exacerbated by urban and peri-urban pumping for city use, which may induce deep infiltration of pollutants into the aquifer system (Morris et al., 2005).

5.3.2 Contamination of water distribution systems

Even where the water supplied from the treatment plant is potable, microbial contamination can occur during or after collection (Gundry et al., 2004). As shown in Figure 5.3, there are numerous potential entry points for contamination of the water supply system. These are described in more detail below.

A significant proportion of disease outbreaks are linked to a deterioration of water quality in the distribution system itself (Ainsworth, 2002) and can be attributed to water contamination caused by breaches in the integrity of the pipework and storage reservoirs (Payment and Robertson, 2004). In addition, there is potential for contamination of public standpipes as well as tankers in the case of vehicular distribution.

Water entering the distribution system may contain free-living amoebae and strains of heterotrophic bacteria, which may colonize distribution systems under favourable conditions and form biofilms (Robertson et al., 2003). However, residual biodegradable organic matter is the most important limiting factor responsible for bacterial regrowth in the water distribution system (Rittmann and Snoeyink, 1984 in Museus and Khan, 2006).

Storage reservoirs

Outbreaks of water-related disease may occur where reservoirs are uncovered or the covers are broken or incomplete, which allows disease vectors to enter into the water system. In addition, uncovered reservoirs may allow for the growth of toxin-forming cyanobacteria and the breeding sites of *A. aegypti* mosquito larvae.

Piped network water supply

According to Robertson et al. (2003), the following are the main causes of contamination in piped distribution systems:

- *Infiltration* – contaminated sub-surface water can enter the distribution system if the following three situations occur simultaneously:
  i)   contaminated water is present in the soil or sub-surface material surrounding the distribution system
  ii) there are pinholes in the piping or joints caused by corrosion, or cracks, and
  iii) there is a low-pressure zone within the system (caused by peak demands, maintenance procedures or power failure).

Figure 5.4 shows a typical example where these situations may occur.

- *Back siphonage* – faecally contaminated surface water can be drawn into the distribution system or storage reservoir as a result of backflow. Backflow occurs if there is simultaneously a temporary negative pressure zone and a physical link between contaminated water and the distribution system. For example, open taps connected to hoses left lying in pools of polluted water may provide this link.
- *Line construction and repair* – water distribution systems can frequently become contaminated after new lines are constructed or after repair of existing supply lines.

5.3.3 Contamination of water in the domestic environment

Regardless of whether or not collected household water is of acceptable microbiological quality as it leaves the treatment plant, it often becomes contaminated with pathogens of faecal origin during transport and storage, due to unhygienic storage and handling practices (Sobsey, 2002). The household environment is a common point for

Public water supply standpipe with water being collected in plastic containers for carrying water and for storage in the home

Rudimentary drainage channel reciving discharges of domestic wastewater and runoff

Illegal water supply connection with poor connections and low pressure

Water supply pipes passing through sewage

Open channel drainage flowing with wastewater

Figure 5.4 **Potential source of contamination of the urban water supply**

Photo: © Jonathan Parkinson

contamination of drinking water, where human health is at risk from the oral ingestion of enteric pathogens. However, the importance of disease transmission in the domestic environment is often underestimated or at worst overlooked. These potential transmission routes are described below.

Distribution systems

Water in household or building water distribution systems can stagnate for long periods, leading to deterioration in the microbial and chemical quality of the water. In particular, Legionella bacteria occupy showerheads and air conditioning units, which are suitable for their survival and growth and may function as amplifiers and/or disseminators of the disease (US-EPA, 1985).

Collection and water storage

In houses where water is not piped directly into the home it is often stored in containers to provide water during times of disruption to the supply. Post-source contamination can occur in storage vessels within households (Gundry et al., 2004). The higher levels of microbial contamination and decreased microbial quality are associated with storage vessels that have wide openings (e.g. buckets and pots) because these are vulnerable to the introduction of hands, cups and dippers that may carry faecal contamination (WHO, 2002). Uncovered storage containers with wide openings are also liable to permit the intrusion of vectors.

5.3.4 Transmission pathways related to drainage

Poor drainage and ponding in urban areas creates conditions that are conducive to the transmission of various diseases. Any receptacle that holds clean rainwater is a potential breeding site for *Aedes* eggs, which results in the proliferation of dengue fever during the rainy season. Disposal of wastewater in the yard may create breeding sites for *Aedes* mosquitoes. The high incidence of dengue is therefore a particular problem where piped water is provided before adequate drainage (Birley and Lock, 1999).

Outbreaks of urban malaria are particularly common during and after the rainy season. The *Anopheles* mosquito lays its eggs in clean, still water that is free from organic pollution, such as pools, which fill with water during rainfall (Kolsky, 1999).

Drainage channels carrying urban wastewater are potential sources of infection and sites for the breeding of insect vectors. Mosquitoes and other pests breed in blocked drains and ponds, spreading diseases such as filariasis. *Culex fatigans* mosquitoes cause some of the most significant environmental health problems as they thrive in polluted water; breeding in pit latrines, septic tanks, sullage pits, drains and in pools containing wastewater effluents (Lines, 2002).

Although runoff from urban surfaces has a low microbial concentration, the risk of contamination increases dramatically when runoff mixes with wastewater and excreta found in foul drains, septic tanks and leach pits, which then disperses pathogens in the environment and increases the risk of infection.

There are many pathogenic bacteria that can cause outbreaks of diarrhoea and gastro-enteritis both during and after storm events. Many of these are caused by ingestion of bacteria from contaminated water. In addition, flooding may disperse parasitic helminth eggs into the environment and provide conditions that are conducive for their survival (see Section 5.2.1).

5.3.5 Wastewater disposal and reuse

Microbiological contamination of natural water bodies used for drinking water abstraction is one of the main concerns associated with wastewater disposal. However, polluted rivers and drainage and irrigation channels may also be used for other activities such as bathing, other recreational activities or washing clothes or food. In these situations, users are at particular risk from various water-related diseases.

In addition, during the dry season, wastewater may be used for irrigating crops in peri-urban areas and for 'greening' of urban areas. As the majority of wastewater is untreated, there are serious health consequences for workers in agriculture as well as for those who consume the produce. There are also public health risks associated with the reuse of wastewater if it is contaminated with industrial wastes containing trace organic compounds and heavy metals.

5.4 STRATEGIES FOR PROTECTING URBAN WATER SYSTEMS FROM CONTAMINATION

This section describes various approaches relating to different components of the urban water system for reducing the transmission of diseases and other health hazards described above.

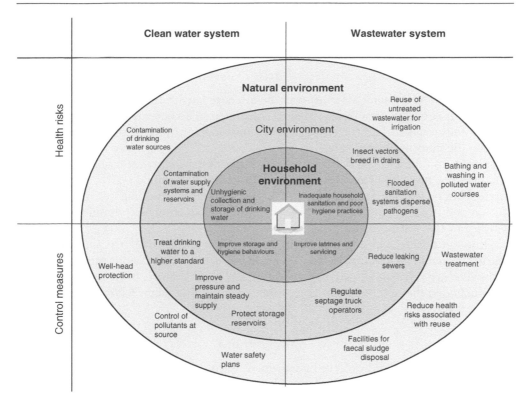

Figure 5.5 Health issues and control measures in the urban water system at different levels

Water treatment technologies fundamental to the protection of public health are presented in Chapter 2. However, there is often an overemphasis on drinking water quality, ignoring other potential transmission routes (as described above).

Figure 5.5 illustrates the range of potential health concerns related to different components of the urban water system in terms of clean water and wastewater systems, but also in relation to the household domain and sources of water supply.

No single intervention can be expected to provide an effective barrier to protect human health. Therefore, public health engineers need to incorporate a range of interventions. This suggests a need for a more holistic approach from the source to the consumer's tap (Medema et al., 2003), based on a multiple-barrier system in which the points of contamination in the water supply chain are targeted simultaneously, as described below.

5.4.1 Protection of water sources

Water source protection involves two steps.

i)   The first step involves the delineation of the boundaries of the wellhead protection area. This requires determination of the direction of groundwater flow, the zone of influence and the recharge area.

ii)   The second step involves a risk assessment, which necessitates the preparation of an inventory of potential contaminant sources (US-EPA, 1993).

5.4.2 Protection of water distribution systems

The distribution system itself must provide a secure barrier from post-treatment contamination as the water is transported to consumers. This is particularly important as disease outbreaks can be traced to cross-connections, in spite of the fact that water leaving the plant is deemed safe.

In order to minimize the risk of regrowth in the system, water entering the distribution network must be microbiologically safe, and ideally it should be free from organic matter. After repairing or replacing existing mains or after installing new mains, disinfection and flushing must be done to prevent the introduction of contaminated soil or debris into the system.

Although disinfection is important, it is not a panacea, particularly as deposits and encrustations in water mains may consume much of the disinfectant and some pathogens are resistant (Ainsworth and Holt, 2004). Therefore, prior to flushing, it is recommended that soil, debris or contaminated water that might have entered the system are removed.

5.4.3 Sanitation and excreta management

Improved sanitation and excreta management is vital for the control of water-related diseases caused by the contamination of water supplies and other transmission routes. The most important primary barrier to transmission of faecal pathogens in the domestic and peri-domestic environment is probably the safe and sanitary disposal of faeces (by the use of latrines, burying, etc.). There needs to be a specific focus on the disposal of children's faeces, which are more dangerous than adult faeces due to higher concentrations of pathogens and because children often defecate indiscriminately – both in private and public spaces.

Even the cheapest sanitation systems can effectively isolate and contain wastes containing excreta, if they are designed appropriately and operated and maintained correctly (Beumer et al., 2002). However, the installation of on-site sanitation is only a partial solution, as no urban sanitation system is completely self-contained. To achieve total sanitation in an urban area, household facilities must be linked to off-site treatment facilities by some form of waste transportation system. Cesspits and septic tanks need to be desludged periodically and the collected waste needs to be removed and transported off-site for treatment.

5.4.4 In-home water storage and domestic hygiene

As the home environment is where many diseases are transmitted, it is important to consider a range of practices that can improve domestic hygiene. Although many of these require some form of technology, they are also closely related to – or in some cases predominantly determined by – behaviour. For example, studies show that the use of covered water containers with narrow openings and dispensing devices such as spouts or taps/spigots are effective in protecting the collected water from contamination by the introduction of microbial contaminants or the intrusion of vectors during storage and household use (Sobsey, 2002). However, residents must also understand why it is important to keep containers covered and this requires a concerted effort on behalf of the institution responsible for promotion.

As well as a wide range of behavioural practices related to sanitation and domestic hygiene, washing hands with soap is of primary importance, as one of the most important routes for transmission of infection is via the hands. According to Beumer et al., (2002), transient microbial contamination picked up onto the hands by contact with a contaminated source can be effectively removed by thorough hand washing with soap and running water.

5.4.5 Drainage

The best way to manage parasite problems is to break their life cycle by destroying their breeding habitats. In practical terms, this means ensuring that soils do not remain wet because of regular flooding. As a result, drainage interventions can result in dramatic reductions in the incidence of infection in communities infested with helminths. For example, in Salvador, Brazil, Moraes et al. (2003) found that a reduction in flooding reduced the incidence of roundworm by a factor of two and hookworm by a factor of three.

The management of runoff is also critically important for the control of mosquito-transmitted diseases, as any amount of standing water can provide a breeding ground for vectors. Strategies to control these diseases need to take into account the nature of the breeding sites of different types of mosquito and the specific nature of the particular mosquito species in the region.

The fundamental objective of drainage for malaria control is to remove ponded water before larvae can mature, in other words, within the duration of the mosquito breeding cycle (one week or less). For health improvements to be achieved it is important that the provision of drainage involves an extension of the drainage system (up to and including the household level), as it is these components of drainage that have a greater impact on the transmission of disease.

The drainage of ponded water is particularly important in the case of dengue control programmes for the control of larval habitats of *Aedes aegypti* mosquitoes, in order to reduce the adult mosquito population. Therefore, strategies to encourage communities to eliminate breeding sites for the *Aedes* mosquitoes, such as water containers, old tyres and other water recipients, are generally more important for the control of dengue than investments in drainage infrastructure (Lloyd, 2003).

According to Beumer et al. (2002), the principal objective should be to avoid the accumulation of sullage, wastewater or clean water in the peri-domestic environment. This may be achieved by adhering to the following guidelines:

- proper apron and drainage arrangements around standpipes/bore-wells
- adequate drainage of rainwater and wastewater from the home
- periodic cleaning and desludging of drains
- in the absence of any proper drainage system, soakage pits should be provided for individual homes or they may be shared by a number of homes, and
- in higher income houses, the accumulation of water in air conditioners and the access of mosquitoes to household drinking water tanks should be prevented.

5.4.6 Wastewater treatment

Conventional wastewater treatment technologies are usually designed to reduce organic and suspended solids loads in order to reduce environmental pollution. The majority tend to be fairly ineffective for pathogen removal (Hillman, 1988), although this is not normally considered to be an objective (unless effluent is reused for irrigation) as the main focus is removal of organic pollutants. In situations where the risk of exposure is high, various types of tertiary treatment technologies need to be used to minimize human exposure to enteric viruses and other pathogens.

The effective disinfection of viruses is inhibited by suspended and colloidal solids in the water. These solids must be removed by an advanced treatment process, such as membrane filtration, prior to disinfection. But achieving such low levels of risk through more advanced wastewater treatment technologies substantially increases costs (Fattal and Shuval, 1999). There are, however, lower-cost technologies, such as maturation ponds, which are effective treatment technologies that can produce an effluent suitable for irrigation (Mara, 1997; Pescod, 1992). Another viable technology is the constructed wetland in which pathogen removal is more efficient than traditional wastewater treatment methods (Williams et al., 1995).

In addition to treatment, there are many procedures to reduce the occupational risks associated with wastewater reuse. These may involve restrictions on the crops grown, as well as various methods used to apply the wastes to the crops to control human exposure (Blumenthal et al., 1989).

5.5 MONITORING THE HEALTH BENEFITS OF URBAN WATER SYSTEMS

Although some technologies are more reliable than others, there will always be uncertainty about the expected health benefits of different interventions. These uncertainties may be compounded by a lack of methodologies for assessing the effects of interventions and the difficulties of choosing the right indicator (Milroy et al., 2001).

Increasing the certainty may be achieved by adopting more intensive sampling and use of more advanced equipment. But the use of these methodologies needs to be weighed against the costs and the time taken to perform the analysis.

The measurement of concentrations of pathogenic organisms in the environment should, in theory, provide an accurate picture of the overall risk to public health. Techniques for pathogen detection are generally complicated, expensive and time-consuming. Therefore, the detection of pathogens in drinking water is often unsuccessful due to the time between contamination and detection of the outbreak.

The most precise measure of health is obtained by monitoring the number of people in a sample human population who are ill at any one time. However, this approach is dependent upon diagnosis of individual cases of infection in the human population. However, this has little benefit from an operational perspective, because by the time the cases of illness are confirmed it will invariably be too late to introduce mitigation strategies.

5.5.1 Conventional approaches towards water quality monitoring

There are two main groups of parameters conventionally used to monitor drinking water quality.

- *Microbiological indicators* – due to the complexities and expense of pathogen monitoring, microbiological indicators are widely used to measure drinking-water quality. These indicators (the most common one is *E. coli*) must be present in sufficient numbers and survive long enough in the environment for measurement to be viable. However, research has shown that pathogens may be present when indicator bacteria are not detected and therefore diseases can still be transmitted by water that meets official standards (Payment, 1998; Medema et al., 2003).
- *Water quality parameters* – parameters such as chlorine residual, conductivity/total dissolved solids and turbidity are commonly used to monitor water quality. Although these do not provide a direct measure of microbiological quality, sudden anomalous changes in these parameters may indicate a problem. In particular, turbidity may indicate the presence of microbial pathogens and serve as a surrogate measure for indicating the risk of microbial contamination (Health Canada, 2003).

5.5.2 Limitations of conventional approaches

The conventional approach to monitoring water quality in a supply system is usually based on testing the drinking water as it leaves the treatment works; sometimes samples may be taken from selected points in the distribution network, or occasionally at the end point of the distribution, in other words, consumer taps. This final option is the most certain from a protection of public health perspective, but it is prohibitively expensive from the point of view of routine quality monitoring.

In addition, water quality monitoring relies on the sampling and enumeration of microorganisms, which is too slow to be used for early warning or control purposes (Payment and Robertson, 2004). Microbiological tests normally take at least a few hours to provide quantitative results of microbial water quality. Therefore, by the time the results are available; the water has been supplied and may already have been consumed. In this case, preventive action is no longer possible.

The other problem relates to the fact that it is normally only possible to carry out tests on a relatively small number of samples, which is statistically insignificant compared to the amount of water produced. Therefore, conclusions drawn about the safety of the water from the results of such sampling are inevitably compromised (Godfrey and Howard, 2004).

5.5.3 Alternative approaches for monitoring of water systems

Ideally, frequent sampling and microbial monitoring at every storage reservoir and service connection throughout the system is required. Although routine microbial quality testing should not be relied upon for operational control, it still plays an important role in verifying that the system is functioning properly. But, as described above, this is usually not feasible. An alternative approach is to introduce real-time monitoring of parameters linked to microbial quality at selected locations throughout the storage and distribution system.

Heavy rainfall may signify the onset of a flood event, often associated with increased incidents of water-borne and water-related diseases. Operational parameters, such as water pressure reduction in distribution networks, can be used to indicate the potential for contamination. With a good knowledge of the system's hydraulics, this approach can be cost-effective and can quickly provide warnings of system failures related to health risks (Godfrey and Howard, 2004).

A Water Safety Plan (WSP) provides a framework for risk mitigation and a management tool to ensure the delivery of safe drinking water. It may be used to provide early warning of potential public health incidents, enabling a planned corrective response.

Water Safety Plans incorporate the approaches towards system monitoring described above, but in addition, periodic sanitary surveys of the storage and distribution system are also recommended. This knowledge can then be used to develop operational plans and identify key priorities for action.

Water safety plans aim to define:

- the hazards that the water supply is exposed to
- the level of risk associated with each hazard
- how each hazard will be controlled
- how the means of control will be monitored
- how the operator can tell if control has been lost
- what actions are required to restore control, and
- how the effectiveness of the whole system can be verified.

By developing a WSP, system managers and operators gain a thorough understanding of their system and the risks that must be managed. The development of a WSP should also identify the additional training and capacity-building initiatives that are required to support and improve the performance of the water supplier in meeting the water safety targets.

REFERENCES

Ainsworth, R.A. 2002. *Water Quality Changes in Piped Distribution Systems*. Geneva: World Health Organization.

Ainsworth, R. and Holt, D. 2004. Precautions during construction and repairs. R. Ainsworth (ed.) *Safe Piped Water: Managing Microbial Water Quality in Piped Distribution Systems*. Chapter 5. London: IWA Publishing.

Beumer, R. et al. (eds) 2002. *Guidelines for Prevention of Infection and Cross Infection in the Domestic Environment: Focus on Home Hygiene Issues in Developing Countries*. 2nd edn. Milano, Italy: Intramed Communications.

Birley, M. and Lock, K. 1999. *Health Impacts of Peri-Urban Natural Resource Development*. Trowbridge, UK: Cromwell Press.

Blumenthal, U.J., Strauss, M., Mara, D.D. and Cairncross, S. 1989. Generalised model of the effect of different control measures in reducing health risks from waste reuse. *Wat. Sci. Tech.*, Vol. 21, pp. 567–77.

Bosch, C., Hommann, K., Sadoff, C. and Travers, L. 2001. Water, sanitation and poverty. *Poverty Reduction Strategy Sourcebook: Water and Sanitation*. Washington DC: World Bank.

CDC. 2005. *Dengue Fact Sheet*. http://www.cdc.gov/ncidod/dvbid/dengue (accessed 23 March 2007).

CDC. 2007. *Leptospirosis Fact Sheet*. www.cdc.gov/ncidod/dbmd/diseaseinfo/leptospirosis_g.htm (accessed 23 March 2007).

Cotton, A. and Saywell, D. 1998. *On-plot Sanitation for Low-income Urban Communities: Guidelines for selection*. Loughborough University: WEDC.

Elankumaran R., Raj Mohan, B. and Madhyastha, M.N. 2003. *Biosorption of Copper from Contaminated Water by Hydrilla verticillata Casp. and Salvinia sp*. Green Pages, July 2003. http://www.eco-web.com/editorial/030717.html (accessed 10 February 2007).

Fattal, B. and Shuval, H. 1999. *A Risk-assessment Method for Evaluating Microbiological Guidelines and Standard for Reuse of Wastewater in Agriculture*. Paper presented at the WHO meeting Harmonized Risk Assessment for Water Related Microbiological Hazards, Stockholm, Sweden, 12–16 September 1999, pp. 1–10. http://www.idrc.ca/en/ev-68330-201-1-DO_gTOPIC.html.

Fields, S. 2004. Global nitrogen: Cycling out of control. *Environ Health Perspect*, July, Vol. 112, No. 10, pp. A556–A563.

Foster, S., Lawrence, A. and Morris, B. 1998. *Groundwater in Urban Development: Assessing Management Needs and Formulating Policy Strategies*. Technical Paper No. 390. Washington DC: World Bank.

Fujiki, H. and Suganuma, M. 1999. Unique features of the okadaic acid activity class of tumor promoters. *J Cancer Res Clin Oncol*, Vol. 125, pp. 150–55.

Godfrey, S. and Howard, G. 2004. *Water Safety Plans (WSP) for Urban Piped Water Supplies in Developing Countries*. Loughborough University, Leicestershire: WEDC.

Gundry, S., Wright, J. and Conroy, R. 2004. A systematic review of the health outcomes related to household water quality in developing countries. *J. Water Health*, Vol. 2, No. 1, IWA Publishing, pp. 1–13.

Health Canada. 2003. *Guidelines for Canadian Drinking Water Quality: Supporting Documentation — Turbidity*. Ottawa, Ontario, Water Quality and Health Bureau, Healthy Environments and Consumer Safety Branch, Health Canada.

Hillman, P.J. 1988. Health aspects of reuse of treated wastewater for irrigation. M.B. Pescod and A. Arar (eds) *Treatment and Use of Sewage Effluent for Irrigation*, Chapter 5. Sevenoaks, UK: Butterworths.

Horan, N. 2003. Protozoa. Mara, D. and Horan, N. (eds) *Handbook of Water and Wastewater Microbiology*. Elsevier, pp. 69–76.

Kolsky, P. 1999. Engineers and urban malaria: part of the solution, or part of the problem. *Environment and Urbanisation*, Vol. 11, No. 1. London: IIED, pp. 159–63

Kuiper-Goodman, T., Falconer, I. and Fitzgerald, J. 1999. Human health aspects. I. Chorus and J. Bartram (eds) *Toxic Cyanobacteria in Water: A guide to their public health consequences, monitoring and management*. UK: Spon Press, pp. 113–54.

Lines, J. 2002. How not to grow mosquitoes in African towns. *Waterlines*, Vol. 20, No. 4, pp. 16–18.

Lloyd, L.S. 2003. *Best Practices for Dengue Prevention and Control in the Americas*. Strategic Report 7. Environmental Health Project, US Agency for International Development, US.

Mara, D. 1997. *Design Manual for Waste Stabilization Ponds in India*. Leeds, UK: Lagoon Technology International.

Mara, D. and Horan, N. (eds) 2003. *Handbook of Water and Wastewater Microbiology*. Elsevier.

Medema, G.J., Payment, P., Dufour, A., Robertson, W., Waite, M., Hunter, P., Kirby, R. and Anderson, Y. 2003. Safe drinking water: an ongoing challenge. A. Dufour, M. Snozzi,

W. Koster, J. Bartram, E. Ronchi and L. Fewtrell (eds) *Assessing Microbial Safety of Drinking Water – Improving approaches and methods*. London: IWA Publishing.

Milroy, A.C., Borja, P.C., Barros F.R. and Barreto, M.L. 2001. Evaluating sanitary quality and classifying urban sectors according to environmental conditions. *Environment & Urbanization*, Vol. 13, No. 1. London: IIED, pp. 235–55.

Moraes, L.R.S., Azevedo Cancio, J., Cairncross, S. and Huttly, S. 2003. Impact of drainage and sewerage on diarrhoea in poor urban areas in Salvador, Brazil. *Trans R Soc Trop Med Hyg.*, Vol. 97, No. 2, pp. 153–58.

Morris, B.L, Darling, W.G., Gooddy, D.C., Litvak, R.G., Neumann I., Nemaltseva, E.J and Poddubnaia, I. 2005. Assessing the extent of induced leakage to an urban aquifer using environmental tracers: an example from Bishkek, capital of Kyrgyzstan, Central Asia. *Hydrogeology Journal*, Vol. 14, No. 1–2, pp. 225–43.

Museus, T.A. and Khan, E. 2006. *A Study of Microbial Regrowth Potential of Water in Fargo, North Dakota and Moorhead*. Technical Report No. ND06 – 01, Dept. of Civil Engineering, North Dakota State University, Minnesota, US.

Payment, P. 1998. Distribution impact on microbial disease. *Water Supply*, Vol. 16, No. 3–4, pp. 113–19.

Payment P. and Robertson W. 2004. The microbiology of piped distribution systems and public health. R. Ainsworth (ed.) *Safe Piped Water: Managing Microbial Water in Piped Distribution Systems*. London, UK: IWA Publishing.

Pescod, M.B. 1992. *Wastewater Treatment and Use in Agriculture*. Food and Agriculture Organization (FAO) Irrigation and Drainage Paper 47. United Nations.

Robertson, W., Standfield, G., Howard, G. and Bartram, J. 2003. Monitoring the quality of drinking water during storage and distribution. *Assessing Microbial Safety of Drinking Water – Improving Approaches and Methods*. London, UK: IWA/Geneva: WHO, OECD, pp. 179–204.

Runnegar, M.T.C. and Falconer, I.R. 1982. The in vivo and in vitro biological effects of the peptide hepatotoxin from the blue-green alga Microcystis aeruginosa. *S Afr J Sci.*, Vol. 78, pp. 363–66.

Sivonen, K and Jones, G. 1999. Cyanobacterial Toxins. I. Chorus and J. Bartram (eds) *Toxic Cyanobacteria in Water: A guide to their public health consequences, monitoring and management*. Geneva: World Health Organization.

Sobsey, M.D. 2002. *Managing Water in the Home: Accelerated Health Gains from Improved Water Supply*. Geneva, World Health Organization.

Stanley, S.L. 2001. Pathophysiology of amoebiasis. *Trends Parasitol*, June, Vol. 17, No. 6, pp. 280–85.

US-EPA. 1985. *Legionella Criteria Document*. United States Environmental Protection Agency, Office of Water, US.

US-EPA. 1993. *Wellhead Protection – A guide for small communities*. EPA 625/R-93/002. Seminar Publication, US Environmental Protection Agency, US.

US-EPA. 1999. *Assessing Health Risks from Pesticides*. EPA-735-F-99-002, Office of Science and Technology, Office of Water, United States Environmental Protection Agency. US.

US-EPA. 2001. *Removal of Endocrine Disruptor Chemicals Using Drinking Water Treatment Processes*. EPA-625/R-00/015, Office of Science and Technology, Office of Water, United States Environmental Protection Agency, US.

WHO. 1997. *Health and Environment in Sustainable Development*. Geneva: WHO.

WHO. 2001. *Arsenic in Drinking Water*. Fact sheet No. 210, Revised May 2001. http://www.who.int/mediacentre/factsheets/fs210/en/index.html (accessed 5 January 2007).

WHO. 2002. *Managing Water in the Home: Accelerated health gains from improved water supply*. Geneva: World Health Organization.

WHO. 2003a. *Emerging Issues in Water and Infectious Disease*. Geneva: World Health Organization.

WHO. 2003b. *Nitrate and Nitrite in Drinking Water – Background document for development of WHO Guidelines for Drinking-water Quality.* WHO/SDE/WSH/04.03/56. Geneva: World Health Organization.

WHO. 2003c. *Lead in Drinking Water – Background document for development of WHO Guidelines for Drinking-water Quality.* WHO/SDE/WSH/03.04/09. Geneva: World Health Organization.

Williams, J., Bahgat, M., May, E., Ford, M. and Butler, J. 1995. Mineralisation and pathogen removal in gravel bed hydroponic constructed wetlands for wastewater treatment. *Wat. Sci. Tech.*, Vol. 32, No. 3. London: IWA Publishing, pp. 49–58.

Chapter 6

Reducing vulnerability to water-related disasters in urban areas of the humid tropics

Eduardo Mario Mendiondo

Engineering School of São Carlos, University of São Paulo, São Carlos, Brazil

6.1 INTRODUCTION

This chapter presents an overview on how to cope with risk and reduce vulnerability to potential disasters in urban areas of humid tropics. The chapter encompasses introductory concepts and ideas on common hazards in urban areas (Section 6.2 to Section 6.6), and strategies for early warning and preparedness (Section 6.7). Section 6.8 covers human resources, in particular, interdisciplinary groups, innovation technology and ethical issues. Finally, a short case study on risk assessment and management, summarizing the strengths and limitations, is presented in Section 6.9. The conclusions in Section 6.10 are not exhaustive but provide avenues of exploration for programmes and policies aiming to enhance the capacity of communities in the face of hazards, vulnerability and exposure to risks.

6.1.1 Terminology

In this section some initial terms are defined for hazard, risk, risk reduction, resilience, vulnerability and adaptive capacity. Detailed definitions are given in Blaikie et al. (1994), Adger (2000), IPCC (2001), UNDP (2004), ISDR (2002, 2005), WMO (2006a, 2006b, 2006c), UN (2006), Lindell et al. (2007), UNESCO (2007a) and Jimenez and Rose (2009).

Hazards are classified into natural, geographical and human-related impacts.

i)   *Natural hazards* include: atmospheric hazards, climatic hazards, hurricanes, blizzards, heat waves, floods, droughts, tornadoes, dust storms, extreme cold, wind storms, tropical storms, avalanches, wildfires, lightening and hail.
ii) *Geographical hazards* include: landslides, earthquakes, volcanoes and tsunamis.
iii) *Human-related (anthropogenic) hazards* include: technological (e.g. toxic, chemical releases, accident spills, nuclear plant accidents and infrastructure collapses) or human-induced hazards, for example, war and terrorism.

A natural disaster is a serious disruption triggered by a natural hazard causing human, material, economic or environmental losses, which exceeds the ability to cope of those affected.

Risk relates to the harmful consequences or expected loss of life, people injured, damage to property or the environment, or disruption to livelihoods and/or economic activity resulting from interactions between natural or human-induced hazards and vulnerable conditions. The term risk, or *how often*, is a composite factor including hazard, vulnerability and exposure. Although alternative definitions exist, the term risk can operationally be expressed by an equation of 'hazard times exposure multiplied by vulnerability'.

Vulnerability, or *how much*, is a combination of a complex and interrelated set of mutually reinforcing and dynamic factors that increase the susceptibility of a community to the impact of hazards. The nature of these factors can be physical, economic, social, political, technical, ecological or institutional. Thus, vulnerability is the degree to which a system is susceptible to, or unable to cope with, adverse effects of natural hazards.

Hence, vulnerability is a function of the character, magnitude, and rate of change and variation to which a system is exposed, its sensitivity, and its adaptive capacity. In practical terms, vulnerability is better addressed in a comparative way. For example, one approach to vulnerability in an urban basin is the relative reduction of the potential water storage of the underlying soil because of the increased amount of paved areas. This example serves well, either for flood vulnerability or for the potential drought index.

Thus, **adaptive capacity**, or *how able to adapt*, is the ability of a system to adjust to sudden disasters, or potential damages, or to extract some kind of advantage from opportunities, or cope with their consequences.

Resilience is the capacity of the urban system or community to resist, change or adapt in order to obtain an acceptable or new level of functioning and structure. This is determined by the degree to which the social system is capable of organizing itself, and its ability to increase its capacity for learning and adaptation, including the capacity to recover from a disaster. There is a need to build up resilience to hazards in society, through a participatory assessment of risks, vulnerabilities and capacities linked to action planning by communities. Urban coping capacity refers to the manner in which urban communities and organizations use urban resources to achieve beneficial ends during the adverse conditions of a disaster phenomenon or process.

Urban disaster **risk reduction** is the systematic development and application of policies, strategies and practices to minimize vulnerabilities, hazards and the unfolding of disaster impacts on urban centres. In accordance with ISDR terminology (ISDR, 2002, 2005), urban disaster risk management is defined as '*the management of legal, administrative decisions and abilities to implement policies, strategies and goals from the society or urban citizens to lessen the impacts of natural and technological hazards*'. This includes all forms of activities, including structural and non-structural measures to avoid (prevention) or to limit (mitigation and preparedness) the adverse effects of hazards (Andjelkovic, 2001).

6.1.2 Urbanization, hazards and disasters

The burden of natural disasters falls most heavily on developing nations, where over 95% of disaster-related deaths occur (IFRC, 2001). The complexity and scale of populations concentrated into urban areas increases the intensity of risk and vulnerability-driven factors. Although smaller cities contribute less pollution to global climate change, they show high levels of internal environmental pollution and risks with regard to ecohydrology (Mendiondo, 2008).

Urbanization presents real challenges both for planning and for the ability of the market to provide basic needs that allow development without creating preventable disaster risks (see Tejada-Guibert and Maksimović, 2001). Urban areas, particularly in developing countries, suffer from rapid changes due to explosive immigration and sprawl. These impose new challenges for coping with risks and disasters for both poor gauged or ungauged basins (see Sivapalan et al., 2003).

Most of the world's population growth between 2000 and 2030 will take place in less-developed regions (Tucci and Bertoni, 2003). One out of every two large cities in the developing world is vulnerable to natural disasters such as floods, severe storms, landslides and earthquakes (Millennium Ecosystem Assessment, 2005).

Natural disasters exert a crucial constraint on development. In doing so, they pose a significant threat to the prospects of achieving the Millennium Development Goals; in particular the target of halving extreme poverty by 2015. Annual average global economic losses associated with disasters increased from US$75.5 billion in the 1960s to US$670 billion in the 1990s (UNDP, 2004). In the period 2000–2006, disasters were large and frequent. Independent estimates show that following the tsunami in late 2004 and hurricane *Katrina*, which affected the south of the United States in 2005, economic losses could be as high as US$2 trillion.

Many of these losses are concentrated in urban areas and fail to adequately capture the impacts of the disaster on poor vulnerable people who often bear the greatest cost in terms of lives, livelihoods and rebuilding their shattered communities and infra-structure. Today, 85% of the people exposed to earthquakes, tropical cyclones, floods and droughts live in countries that have either a medium or low Human Development Index (HDI). During the decade 2000–2010, population increases are predicted to occur most rapidly in urban areas in the countries of Africa, Asia, and Latin America and the Caribbean, with more than half of the world's population becoming urbanized by the period 2007–2010 (UNDP, 2004). It is important to state that in urban areas everyday risks have cumulative effects which are frequently converted into disasters especially for more vulnerable communities. Box 6.1 gives a short discussion of this issue for African urban areas.

Impacts from disasters have, however, a different context. For every ten deaths from disaster, five are due to war and four are due to famine (Carmin, 2005). Though natural disasters may seem less important, they still have a significant impact. Every day, at least 184 people die from a natural disaster, and approximately 3,000 injuries are associated with every death by natural disaster. Between the 1960s and the 1990s, the global impact of recovery from disasters grew with a constant rate of US$20 billion a year. But these numbers are known to be under-calculated (see, for example, UNDP, 2004; ISDR, 2002, 2005). In addition, transient populations are often not counted and some properties are undervalued in these estimates. The real losses associated with disasters from natural hazards are therefore greater than these estimates.

However, progressive thinking about disaster planning and mitigation is in constant evolution. On the one hand, cost-based losses based upon utilitarian approaches are

---

Box 6.1 From everyday hazards to disasters: risk accumulation in African urban areas

Many disasters take place in urban areas, affecting millions of urban people each year through loss of life, serious injury and loss of assets and livelihoods. Poorer urban groups are generally the most affected. The impact of these disasters and their contribution to poverty are under-estimated, as is the extent to which rapidly growing and poorly-managed urban development increases the risks.

Urban specialists often do not view disasters and disaster prevention as falling within their remit. Moreover, few national and international disaster agencies have worked with urban governments and community organizations to identify and act on the urban processes that cause the accumulation of disaster risk in urban areas.

These discussions were summarized at a workshop funded by UNDP on the links between disasters and urban development in Africa, which highlighted the underestimation of the number and scale of urban disasters, and the lack of attention to the role of urban governance. The difficulties of instigating action in Africa were noted, since the region's problems are still perceived as 'rural' by disaster specialists, even though two-fifths of the African population live in urban areas.

This emphasizes the need for an understanding of risk that encompasses events ranging from disasters to everyday hazards and which understands the links between them – in particular, how identifying and acting on risks from 'small' disasters can reduce risks from larger ones. It also stresses the importance of integrating such an understanding into poverty reduction strategies.

Source: Bull-Kamanga et al. (2003)

---

an integral part of vulnerability assessments (see Box 6.2), but increasingly non-utilitarian approaches incorporate additional elements such as those that take into account aesthetic and cultural issues (Millennium Ecosystem Assessment, 2005).

6.1.3 Need for lessons to be learned

According to figures from UNESCO's Division of Basic and Engineering Sciences, for every US$100 spent by the international community on risk and disasters, US$96 go towards emergency relief and reconstruction, and only US$4 on prevention. Yet, each dollar invested in urban water risk prevention reduces by up to US$25 the losses incurred in the case of natural disasters. Box 6.3 gives a summary of lessons learned to reduce vulnerability to urban water disasters.

The situation related to the risk management of water hazards and ways of coping with potential urban water security disasters is unique. Moreover, the hydrologic risks associated with disasters are expected to increase with exploding demand and exposure in urban areas assessed from global scenarios of climate change.

The handling of these risks necessitates using approaches that involve stakeholders to achieve water security. The failure of policymakers to promptly respond to solvable water crises results not only in massive unnecessary losses in urban areas, but also in a vicious cycle of poverty from non-continuous policies. This cycle is caused by the non-prevention of potential water risks, in other words, floods, droughts, water-conflicts, and so on.

When such risks are not well-prevented they may become social disasters, related to both unwise water policies and lack of preparedness steps. These non-prevented risks

Box 6.2 Cost-based losses in urban areas and the disaster vulnerability perception

Disaster losses are categorized as direct costs, indirect costs and secondary effects.

Direct costs are physical damages, including those to productive capital and stocks (industrial plants, standing crops, warehouses, etc.), economic infrastructure (roads, electricity supplies, etc.) and social infrastructure (homes, schools etc).

Indirect costs are related to downstream disruptions to the flow of goods and services, in other words, lower output from damaged or destroyed assets and infrastructure and the loss of earnings as income opportunities are disrupted.

Disruption of the provision of basic services, such as telecommunications or water supply, for instance, can have far-reaching implications. Indirect costs also include the costs of both medical expenses and lost productivity arising from increased incidence of disease, injury and death. However, gross indirect costs are also partly offset by the positive downstream effects of reha-bilitation and reconstruction efforts, such as increased activity in the construction industry.

Finally, secondary effects comprise short and long-term impacts of a disaster on overall econ-omy and socioeconomic conditions, for example, fiscal and monetary performance, levels of household and national indebtedness, the distribution of income and scale and incidence of poverty, and the effects of relocating or restructuring elements of the economy or workforce.

For example, the Brazilian heavy rains in early 2004 provoked regional floods and the impact of which was estimated at about US$3.5 billion, including direct, indirect and secondary costs. Of these figures, 65% affected urban areas. Notwithstanding, governments advocated assistance based on their own perceptions of 'the disaster' and implemented official emergency plans amounting to US$50 million.

This enormous gap between cost-based losses and disaster vulnerability perception should form the basis of a discussion on policies to target greater efficiency between real impacts and public awareness.

Source: Integrated River Basin Group, Dept Hydraulics & Sanitation, EESC/USP, www.shs.eesc.usp.br

---

Box 6.3 Lessons learned to reduce vulnerability to urban water disasters

The following list includes some of the main factors that can help reduce vulnerability in urban areas.

- *Collaboration and consultation* are essential for identifying needs and gaps, learning about a community's learning styles, and developing ongoing support for their projects.
- *Participation of affected communities*, as natural disaster-affected communities want to be involved in projects that will lessen the impact of future natural disasters. They must be viewed as a valuable resource rather than passive recipients of donor aid.
- *Language barriers.* It is important to use local languages effectively in order to deliver natu-ral disaster preparedness messages. However, the universality of a project may be lost if a lack of disaster preparedness terminology in a local language inhibits effective disaster pre-paredness communication in urban areas.
- *Culture and religion* require sensitivity in order to develop innovative approaches to pro-moting communication and understanding where certain cultural beliefs and practices may present obstacles to natural disaster preparedness.
- *Government and programmatic support* are crucial to developing sustainable, ongoing com-mitment to local stakeholder initiatives at local district levels, through policy, financial or coordination efforts.

Source: adapted from UNESCO, 2007a

foster a 'water-poverty-cycle': a circuit of post-event, reactive and inefficient activities that irreversibly affect communities in metropolitan flood-prone areas (Mendiondo, 2005a).

In the 1990s, with the International Decade for Natural Disaster Reduction, the focus on disaster management was related primarily towards issues of poverty, globalization and mega-cities. However, an emerging view is focusing on thinking about disasters in relation to their socio-political contexts, and seeking ways to support the development of more resilient systems. For this to be achieved, there is a need for both the public as well as professionals to develop a better understanding of the complex factors involved (Goldenfum and Tucci, 2005).

6.2 STORMS, FLOODS AND CYCLONES

Storms, floods and cyclones are mainly associated with water disasters in subtropical urban areas. In the case of South America, natural hazards have destructive and sometimes lethal impacts, particularly in urban areas (Mendiondo and Valdes, 2002). Of the total number of registered events between 1900 and 1998, 66% were weather or climate-related: 34% due to floods, 5% due to droughts.

Over the last 100 years, millions of people have been affected as their housing, sources of income and communities have been damaged or destroyed by extreme floods. Hurricanes provoke significant losses in urban areas. Estimates of damages caused by Hurricane *Mitch* in Central America totalled US$6 billion in 1998, the equivalent of 16% of the average GDP (Gross Domestic Product), 66% of exports, 96.5% of gross fixed capital formation and 37.2% of the total external debt. Before *Mitch*, Central America's projected GDP growth for the years 1999 to 2003 was 4.3% per year. Subsequently, it was estimated at only 3.5%, or 1.2 points lower than it would have been had the hurricane not occured (Mendiondo and Valdes, 2002). As measured by GDP, the losses in Argentina in the 1991 floods were US$1.2b (~1% GDP), and in Honduras the 1998 *Mitch* hurricane caused losses of approximately US$4b (~100% GDP).

Box 6.4 summarizes the change in policy in South America over the past couple of decades from disaster mitigation to urban early-warning systems.

6.3 WATER SHORTAGE AND DROUGHTS

Although the humid tropics usually suffer from water extremes such as floods and hurricanes and interdecadal climate variability, potential global warming is likely to increase natural disasters impacts. As a result, water shortages will appear in cities previously considered as located in humid regions. In theory, a drought is a recurrent natural feature, progressively emerging near urban centres. It results from the combination of lack of precipitation over an extended period and the growing water demands of society.

Depending on the likely impact, the phenomenon can be categorized in several ways, for example, meteorological, hydrological, agricultural or sector-driven. The spatial extent of a drought is much greater than for any other hazard, and is not limited to a city or basin or by political boundaries. For instance, trade in goods from increasing semi-arid regions will affect cities of the humid regions of the same

---

Box 6.4 From disaster mitigation to urban early-warning systems in South America

The accumulated losses from the South American floods in the period 1995 to 2004 were approximately US$25 billion, including immediate repair and operation and maintenance (O&M) costs. Post-flood restoration plans transfer annual debts, with rates of US$30 to 3,500 per capita, to populations living in small human settlements and metropolitan areas, respectively.

In this way, South American countries are affected annually by a 2% to 5% decrease in GDP caused by non-prevented floods, which are converted to social disasters – the so called 'flood-poverty-cycle'. Preliminary estimates up to the year 2050 outline the possibility of reducing costs at a relative annual rate of US$150 per capita through wise flood management in South America, and reducing the flood-poverty-cycle by an order of magnitude of fifty times with appropriate flood strategies and new public-private partnerships.

Until 2010, there will be a net positive increase in O&M flood losses of about US$2 billion a year, either because of lack of prevention or contingency flood plans. Alternative scenarios show that O&M costs could reduce to US$0.5 billion a year until 2020 if early warning systems are included in river committee plans, with an increase in the monitoring and forecast of impacts in ungauged basins.

The total 'flood market' in South America is estimated at about US$80 billion a year and is expected to increase until 2075, converting reactive policies to proactive ones. A transition phase in South American urban river basins is foreseeable, as concepts of flood disaster mitigation are increasingly replaced by visions of early-warning systems and transfer-risk devices through flood insurance.

Source: Mendiondo (2005a, 2005b)

---

transboundary river basin. Its impact is difficult to quantify, but will slowly accumulate over years and vary according to the society and regions concerned.

Long-lasting droughts lead to the degradation of soils, plant and animal habitats, as well as social disruption. Box 6.5 has a summary of water availability and vulnerability of ecosystems and society from an interdisciplinary and stakeholder approach.

Urban drought severity is dependent not only on the duration, intensity and geographical extent of precipitation deficiency across river basins, but also on urban water demands for supply. The particular features of urban droughts are relevant because the annual effects of drought tend to accumulate slowly over time and may linger for years before longer-term strategic urban plans can be developed.

In addition, the end of urban drought can be difficult to determine, while the impacts of drought are less obvious than the damage that results from other natural hazards. Araujo et al. (2004), working with results from municipalities of northeast Brazil (see Box 6.5), propose a new index for predicting droughts based on long-term scenarios to overcome urban drought severity and to reduce the vulnerability of local communities to migration.

The vulnerability of cities to interruptions to water supply is even more apparent. Water rationing is one potential option, but in reality, given the poor operational logistics of city water authorities, in practical terms this means that fewer and fewer residents receive water services. Competition for water use is intensifying in urban areas, and treatment costs may increase sharply.

Water shortage and potential droughts are also relevant to urban areas of the humid tropics. One vital task for water managers is therefore to reduce water loss in urban

---

Box 6.5 Water availability and vulnerability of ecosystems and society: the WAVES project

Working at 332 municipalities in northeast Brazil, the project WAVES – Water Availability and Vulnerability of Ecosystems and Society – introduced policy scenarios among interdisciplinary research teams, water practitioners and stakeholders. In order to address regional water impacts from global changes, land use and water-related scenarios across about 400,000 km^2 of water scarcity area were derived for the period 2000–2025. These scenarios had the potential to support strategic planning by the water and agriculture authorities to mitigate droughts and introduce the concepts of living with water scarcity risks in urban centres and peri-urban settlements.

Qualitative-quantitative reference scenarios up to the year 2025 were generated by developing storylines, quantifying the driving forces and applying simulation models. This approach has been adapted from one scenario of the Intergovernmental Panel on Climate Change (IPCC), developed to derive global scenarios on greenhouse gas emissions. The WAVES project introduced policy workshops with stakeholders and decision-makers in order to refine former model outcomes and partial hypotheses, and to better guide the potential adaptation of local policies to cope with water scarcity and droughts. Related models on disaggregation of GCMs, migration, water quality, in-farm management and water use modeling were integrated into the WAVES project.

Some lessons learned pointed to the necessity of having not only indicators and integrated models, but also new adaptive policies or hydrosolidarity protocols that propose a comprehensive learning process for urban water-users, for whom water scarcity was inherent in the past and should be decisive in the future. Further capacity-building from the WAVES experiment helped to create new water-resilient visions, coping strategies and feasible goals for other worldwide research groups to reduce the vulnerability of ecosystems and society.

Source: Center for Environmental Systems Research, University of Kassel, www.usf.uni-kassel.de/waves/

---

areas. In many African, Asian and South American towns, water losses of up to half of production are caused by leakage and illegal connections. However, institutional bottlenecks combined with poor operational practices make this a cumbersome task.

6.4 LANDSLIDES AND MUDSLIDES

Landslides and mudflows result from interactions between hydrological and geological processes. These geological processes are either triggered by hydrological conditions or may occur independently. Landslides and mudflows can also be caused by heavy rainfall, when soils become saturated and the stability of slopes are no longer maintained. In Venezuela, the 1999 mudflows caused US$3.2 billion of damage (3.3% of GDP).

The amount and/or intensity of rainfall necessary to trigger an event also depends on the soil properties and the steepness of land slopes. Therefore, general weather and rainfall forecasts have to be transformed into warnings at the local level and preemptive evacuation plans put into effect.

These damaging phenomena are characterized by high density and rapid movement and thus have the potential to destroy buildings and other infrastructure in their path. Landslides occur only within well-defined geological and topographic areas. Hazard mapping can be carried out to delineate such high-risk areas based on terrain analysis. The traces of old slides and mudflows can often be used for this purpose.

Box 6.6 Managing policies for urban users at the Asian Disaster Preparedness Centre

The Asian Disaster Preparedness Centre (ADPC), established in 1986, is a regional, inter-governmental, non-profit organization. Its mandate is to promote safer cities and sustainable development through reduction in the impact of disasters in response to the needs of countries and communities in the Asia-Pacific region. It does this by raising awareness, helping to establish and strengthen sustainable institutional mechanisms, enhancing knowledge and skills, and facilitating the exchange of information, experience and expertise.

The ADPC develops disaster risk management programmes and projects by providing technical and professional services for the formulation of national disaster management policies. It facilitates the development of institutional mechanisms to support disaster risk reduction, capacity-building of disaster management institutions, programme design for comprehensive disaster risk management, post-disaster assessment, public health and emergency management, land-use planning, disaster-resistant construction, and the planning of immediate relief response and subsequent rehabilitation activities.

In twenty years, the ADPC has responded to the paradigm shift in disaster management, readily and actively adjusting its operational strengths to address evolving developments, and structuring its technical focus on disaster risk management, particularly in urban areas. This comprehensive approach is further reinforced by ensuring a more prepared and aware community through education and awareness-raising initiatives with schools and colleges. The ADPC, through this partnership with UNESCO, strives to promote a culture of preparedness and prevention by promoting and supporting the mainstreaming of education of disaster risk reduction of urban and peri-urban communities.

6.5 TSUNAMIS

The growing urban population of certain humid tropics has provoked expansion from country to coastline areas where tsunamis can occur. Tsunamis are generated by earthquakes, which create long, fast-moving waves that can be 10 to 20 m high by the time they reach the shoreline. A total of 158,551 deaths were associated with tsunamis and earthquakes around the world between 1980 and 2000 (UNESCO, 2007a). The devastating capacity of a tsunami was tragically demonstrated on 26 December 2004 in South and Southeast Asia. Some initiatives like the Asian Disaster Preparedness Center (see Box 6.6) envisage community-based assistance to cope with tsunami risks to human settlements and urban areas.

6.6 HUMAN-RELATED HAZARDS

Many built structures in urban areas of the humid tropics have been designed to cope with water risks so as to reduce vulnerability to disasters, and to make the best use of water resources. Decision-makers face increasing problems, however, from assessments of potential cancer risks from *trihalomethanes* in the water supply, deterioration of drinking-water sources through leaching of organic matter, and latent risks from *helminth* eggs in wastewater and sludge (UNESCO, 2007b).

A new paradigm on managing risks in urban areas has been provoked by the advent of several challenging issues, including recent contributions to biological, chemical and physical risks and their relation to urban water security, with a special focus on human and ecosystem health (see Thuo, 2001). Any technical structure or system may fail, increasing risks due to loss of efficiency, which could in turn intensify impacts from risks of extreme water-related events. For example, service failure on the part of a water distribution company can cause a hazard higher than any drought due to the lower resilience of urban dwellers who lack both preparedness and equipment to recycle part of by-products from water treatment, i.e. removal of trihalomethanes from chlorination.

In such cases, new technologies need to be linked and managed at the local level (e.g. reuse of water, cistern storage, infiltration trenches and green roofs). New efficient flushing toilets, intelligent bathroom and parallel domestic storage system technology are viewed as localized devices for reducing the failure of the urban water cycle at home.

However, constraints on the reuse of grey or wastewater introduce a different type of risk, related to the handling of and exposure to pathogenic agents. The risk of spreading diseases, such as dengue, transmitted by the mosquito *Aedes aegypti* whose larvae are deposited in clean water tanks, or malaria whose principal mode of spread is the bites of the female *Anopheles sp* mosquito, is increasing in the humid tropics (see also Chapter 4, with detailed topics on health risks). During 2008, dengue surges in the state of Rio de Janeiro, Brazil, spread out of control in urban areas and were categorized as an epidemic health disaster.

Exposure to microorganisms in water is therefore a crucial factor for all alternatives designed to mitigate human related-hazards. It has received positive attention in policies and regulations instituted by authorities. However, soil contamination with greywaters, which contain different types of pollutants and leads to a steady deterioration of groundwater quality, is a cumulative risk that is, as yet, not adequately discussed by urban decision-makers.

Potential options for low-income communities include reuse of greywater and storage of rainwater for toilet flushing, reuse of water for irrigation purposes, and even carbon assimilation due to fertilization (Galavoti et al., 2007). The replication of initiatives to reduce water vulnerability in urban centres of the humid tropics through decentralized, cheaper solutions also appears feasible under integrated water resources schemes, which highlight early warning and preparedness. However, both require specially tailored implementation strategies to ensure successful uptake and sustainable operation.

6.7   EARLY WARNING AND PREPAREDNESS

Early warning systems, water proofing of dwellings at risk, and indigenous or native knowledge related to predicting hazards all form part of the adaptive capacity to mitigate urban disasters (UNDP, 2004). Both large-scale and localized integrated water resources schemes pose risks, which need to be reliably quantified. But those urban communities with the least resources are more vulnerable and have the least capacity to adapt to hazards because of greater exposure and lack of infrastructure.

Preparedness and perceptions are central to management. Preparedness consists of preventive and precautionary measures to prepare for an event before it occurs. It aims to minimize the impact of development activities that accentuate the magnitude of hazards, reduce exposure to natural hazards, and minimize socioeconomic vulnerability for people and material assets exposed to these hazards. It also deals with the issues of increasing risk and risk aversion (see Diamond and Stiglitz, 1974).

Prevention deals with long-term planning, and is incorporated into the development process. Prevention relates to 'risk perceptions' of local communities and is deeply rooted in the social, cultural and religious ethos of a society itself. Risk management measures constitute the full complement of actions implemented prior to, during, and after the hazard event.

6.7.1 Supporting agencies

The design of preparedness programmes and plans for disaster response are crucial for the community (Lindell et al., 2007). Other mitigating actions to support communities include training and capacity-building to reduce physical vulnerability, ameliorate the economy and strengthen the social structure. These actions can be undertaken at individual, community and state levels. Non-governmental organizations, voluntary and socio-cultural organizations may also play an important role in this respect.

6.7.2 Challenging issues of insurance for water disasters

To build more resilient communities, it is essential to take an integrated approach to management, linking land and water uses, hazard risks, socioeconomic development and the protection of natural ecosystems through institutional framework and public participation. More challenging issues are related to non-structural measures, such as, land zoning, participatory empowerment and risk-transfer through insurance devices (Andjelkovic, 2001; Crichton, 2008; UNEP FI, 2009).

Figure 6.1 outlines three simulations, from fifty-year runs of an annual flood insurance fund, performed through an insurance model in a Brazilian urban catchment containing areas prone to frequent flooding. The results of the insurance fund are

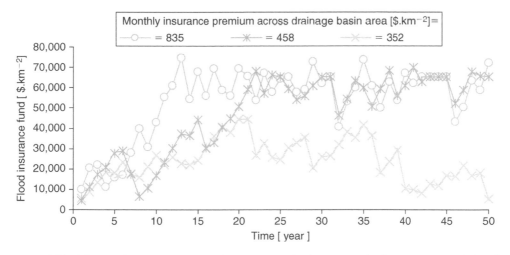

Figure 6.1 Model runs of the performance of an insurance fund, at nominal values, to cope with flood hazards in an urban subtropical basin (monthly premiums are depicted in monetary values per drainage area of the basin)

Source: Mendiondo et al., 2005b

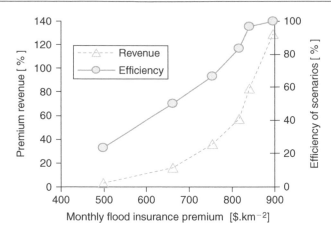

Figure 6.2 **Average revenue (left ordinates) and time efficiency (right ordinates) according to insurance premium to reduce flood vulnerability in urban subtropical basin**

Source: Mendiondo et al., 2005

expressed in annual US$ per square kilometre and are based on a combination of hydrological, hydraulic and economic-based sub-components (Mendiondo et al., 2005). A difference in the monthly premium per square kilometre produces variations in the overall inter-annual insurance fund: lowering the premium results in a less sustainable insurance fund. Consequently, Figure 6.2 illustrates flood insurance as a risk transfer device for an urban basin, with revenues and efficiency.

Figure 6.2 comprises annual-based simulations through runs of fifty consecutive years, through hydrologic damage curves, surveyed locally; stochastically-based replications of thirty scenarios of the same insurance coverage; and under six different types of economic insurance coverage (Mendiondo et al., 2005).

Progressive urban development and climate variability – as well as sensitive damage curves – introduce particular elements, which can be incorporated into the insurance models of Figures 6.1 and 6.2 to adapt them to different urban realities. These elements include vulnerability curves, risk exposure within prone areas monitored by web-GIS, and time-delay tracking of vulnerable people, for example, children or elderly people.

The risk-transfer model can encompass alternative approaches of assessing resiliency and vulnerability. In Figure 6.2, premium revenue (%) highlights the difference between the initial, non-optimized premium and the final, optimal premium. This acts as an alternative index to emulate the resilience capacity of insurance as a risk-transfer device. Otherwise, the right ordinates of Figure 6.2, relating to efficiency, show the relative number of favourable long-term scenarios at which the insurance fund is self-sustainable. In these terms, efficiency is an alternative indicator to express how to cope with disaster vulnerability in an urban basin as a whole.

Where insurance is required, insurers may be approached by developers in advance of an application for a development permit. A development is unlikely to proceed if insurance is refused or premiums are too high owing to the risk of flooding. In the United States, local governments are given the choice of participating in the National Flood Insurance Program (NFIP) to ensure that adequate flood management plans are in place. States are thereby eligible for federal aid following a flood event, if they maintain floodplain control programmes. A number of financial incentives are in place to encourage participants to exceed the required standards.

An alternative method has been to prohibit regulated lenders from providing mortgages for development in certain areas. It remains to be seen if changes are made to this system following hurricane *Katrina* in 2005.

Although flood insurance is seldom mandatory, problems can occur in regions where insurance is optional. In such situations, insurance is bought solely by those who are aware that they are at risk and have experienced flooding in the past. Consequently, the premiums can become more expensive, further discouraging others from buying insurance. Unfortunately, in the developing world, insurance is not an option for the majority of people, who shoulder the costs themselves, the availability of government aid being uncertain in many cases.

6.8 HUMAN RESOURCES

Lindell et al. (2007) identify the hazards, emergencies and disasters that can have a potential impact on a community. Human resources and capacity-building primarily encompass interdisciplinary groups and innovation in technology. Interdisciplinary groups are crucial to ensure the use of the right language in the right company. Innovation technology calls on the experience of previous water projects, looking at ways in which the shortcomings of older works could be newly adapted to manage challenging issues, in other words, to reduce vulnerability to water-related disasters in urban areas. In addition, every innovation, whether necessary or mandatory, has to attain agreement from new audiences, such as new stakeholders, in order to acquire and maintain widespread acceptance.

6.9 CASES OF RISK ASSESSMENT AND MANAGEMENT

In the case of several alternatives, risk assessment and risk management can be combined. The following case of urban flood risk evaluation is a methodological example that looks at the steps of assessment and management, through operational equations and costs applied to a real case study of flooding in an urban area of the humid tropics. More details are explained in Mendiondo et al. (2005) and Mendiondo (2005b). In the following sections both risk assessment and risk management are discussed in brief in order to depict how previous concepts should be operationally integrated with indicators for policymakers.

6.9.1 Risk assessment

From a hypothetical perspective, imagine an urban basin which was a formerly undeveloped basin. Its potential water storage in the underlying soil was S_{past}, which has decreased with time, $\Delta S = S_{present} - S_{past}$, where $S_{present}$ is the present water soil water storage after growing urbanization. Both S_{past} and $S_{present}$ can be assessed through traditional runoff approaches related to the level of basin imperviousness. The ratio $\Delta S \div S_{past}$ is the first measure of the basin's vulnerability to the urbanization process, which provokes increasing potential runoff.

To outline the relative time-related vulnerability, suppose that the time a more vulnerable person needs to walk through a flood-prone valley, t_v, is divided by the duration a normal, less vulnerable, person needs to walk the same distance, t_f.

The risk exposure factor has two components. First, the ratio of the flood-prone area with regard to the return period hazard, $A_{f,Tr}$, and the total basin area, A_b. Second, the number of people living in flood-prone areas, $n_{f,Tr}$, divided by the total population of the basin, n_b, expresses the relative exposure of the total basin population to that hazard level. Note that both flood-prone areas of potential inundation, $A_{f,Tr}$, and the people affected, $n_{f,Tr}$, could vary according to the magnitude and return period of the flood hazard, Tr_f. The hazard ratio could be assessed by the quotient of return periods, that is, of the hazard Tr_f itself relative to the average return period T^* of the design bank-full condition of the main river channel at that urban basin.

6.9.2 Risk management

Risk management steps involve the coping capacity of operational actions before, during and after hazards occur. These actions show, respectively:

- the ratio of a basin's time of concentration, t_{con}, to the early-warning prediction time, $t_{early-w}$
- the ratio between the time of rescue teams who attend the emergency under hazard impacts, t_{rescue}, and the time of concentration of the basin, t_{con}, and
- the traditional expression of risk when sizing structural measures from a return period and lifetime of the structure, as a mitigation measure for potential rebuilding.

In this case, the full river channel capacity is sized for a specific hazard return period, T^*, throughout an operation and maintenance lifetime, N, minus the respective amortization time j. It is worth noting that t_{con} is related to the basin population density $(n_b A_b^{-1})$ which varies with time, and $t_{early-w}$ and t_{rescue} are related to technological and institutional capacity-building, which can vary according to planning policy scenarios across decades.

The costs of risk management in urban humid areas are different and vary according to the management step (Mendiondo, 2005b). The evaluation of cost-risk units gives $Unit_{Cost}$, expressed as \$ m^{-2} inhab^{-1}, assessed on total costs that affect housing areas and people living in flood-prone areas.

The actions and costs associated are different. In general, early warning actions cost approximately 1 unit of risk ($Unit_{Cost} \approx 1$), whereas contingency plans increase to 4 units and hard control towards an average value of 25 units (Mendiondo, 2005b). The total risk cost can be evaluated by multiplying $Risk$ expression by $Unit_{Cost}$, in a cumulative approach through time. In short, risk assessment and risk management are given as a set of variables, which give $Risk\{ \cdot \}$ as a composite indicator expressed by a sequence of factors of hazard, exposure, vulnerability and management steps, before, during and after the occurrence of the hazard episode. These factors can take the form of several combinations of non-dimensional variables, multiplying risk units as time-progressive indicators of risk level and potential disasters accordingly. One expression of this composite risk indicator is shown as follows:

$$
Risk \left\{ \begin{array}{c} \overbrace{\underbrace{\left(\dfrac{Tr_f}{T^*}\right)_j}_{hazard}, \quad \underbrace{\left(\dfrac{A_{f,Tr}}{A_b} \cdot \dfrac{n_{f,Tr}}{n_b}\right)_j}_{exposure}, \quad \underbrace{\left(\left|\dfrac{\Delta_s}{S_{past}}\right| \cdot \dfrac{t_v}{t_f}\right)_j}_{vulnerability}}^{risk-assessment(components)}, \\[4ex] \overbrace{\underbrace{\left(\dfrac{t_{conc}}{t_{early-w}}\right)_j}_{before(prediction)}, \quad \underbrace{\left(\dfrac{t_{rescue}}{t_{conc}}\right)}_{during(contingency)}, \quad \underbrace{\left(1-\left(1-\dfrac{1}{T^*}\right)^{N-j}\right)}_{after(reconstruction)}}^{risk-management(steps)} \end{array} \right\}
$$

where sub-index f indicates a random *flood* hazard and j refers to an annual time step. Other sub-indexes, such as d for droughts, *land* for landslides, and so on, can be performed using this equation with corresponding adaptation of new variables for specific hazards.

Note also that the set of variables is simple because the expression looks, for operational purposes, to reduce risk with a plurality of approaches relating to inter-disciplinary groups or stakeholder sectors, according to each term: that is, with urban planning ('exposure'), hydrology and social/health sciences ('vulnerability'), meteorologists ('before'), emergency task forces ('during') and the construction sector ('after'). For this reason, the set of variables should not be approached as unique, but as a measure towards an integral approach to risk mitigation. However, when specific problems are to be solved, science-based risk equations (Simonovic, 2005) and community-based approaches (Costea and Felicio, 2005; Lindell et al., 2007) are preferable.

6.9.3 Strengths and limitations: a short example

To illustrate the strengths and limitations of the former approach, a short example is shown for a prospective planning period of 2000 to 2050, with a common insurance model. This example is related to the same basin as in Figures 6.1 and 6.2, and is adapted from the insurance model presented by Mendiondo et al. (2005). A subtropical urban catchment with a drainage area of $13\,km^2$ is tested under different policies with flood-prone areas affected by return period hazards. Risk assessment and risk management are estimated in a multiplicative way, constituting a preliminary option for discussions with decision-makers. Costs are taken from risk management tables for flood hazards (Mendiondo, 2005b). Very strong hazards with return periods higher than twenty years ($T^* > 20$) are simulated to occur in 2028 and 2039. The predicted 2036 hazard is expected to be catastrophic because its return period would be

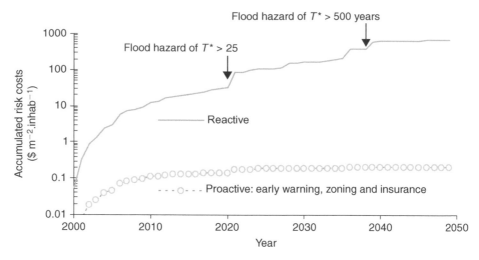

Figure 6.3 Simulation of accumulated nominal costs from two policy scenarios of risk management to cope with flood hazards and growing urbanization for an urban subtropical catchment drainage area of 13 km². Proactive policies have early warning systems, land zoning of flood-prone areas and insurance for risk-transfer

Source: adapted from Mendiondo et al., 2005

$T^* > 500$ years. Either a reactive or a proactive policy is established for flood risk management. Proactive policies envisage land zoning, early-warning and insurance mechanisms for risk transfer. Figure 6.3 shows the cumulative costs of flood risk management for the period 2000–2050. Note that the resiliency capacity of the proactive policy is high; accordingly, the accumulated costs of risk management of reactive policies are relatively ten times higher for a medium hazard ($T^* > 20$) and about 100 times higher for the highest hazard ($T^* > 500$ years).

Figure 6.3 lacks the dynamic situations of real-time floods. For dynamic situations, a hazard factor ($T_{rf} \div T^*$) could be substituted from hydrodynamic forces during the flood wave passage of a hazard of the corresponding return period. Such a factor could be replaced, for example, by a ratio of inertial factors of average flow velocity times water level, as $v \cdot h$. Thus the numerator would refer to values on the floodplain from hazard event $(v \cdot h)_{Trf}$, and the denominator would also relate to the vulnerability threshold to which a submerged body would hold the inertial force of the flow, $(v \cdot h)_{vul}$.

The strength of this approach is that it could be introduced into urban policies and strategies to reduce risks and their costs through a kind of signboard or flood-semaphore. This would be of use to disaster specialists, public agents and planners, strategically approaching risk in terms of relative coefficients in order to cooperate with users and stakeholders (see further discussions in Andjelkovic, 2003; WMO, 2006a; WMO, 2006b; UNESCO, 2007a; UNESCO, 2007b).

6.10 CONCLUSIONS

This chapter briefly explained the concepts, strategies and practices used to address reduction of vulnerability to hazards in urban areas, focusing on certain components of the water cycle. Several authoritative papers, guidelines and programmes here cited are already available worldwide, and provide valuable scientific insights and practical perspectives for stakeholders.

Local values, like hydrosolidarity and traditional local knowledge, play a crucial role in reducing the impact of hazards in urban areas. Community-based management is vital in order to avoid converting natural hazards into catastrophic, but preventable social disasters. Thus, cycles of poverty related to unmanaged hazards should be reduced in the medium or long-term.

Non-structural measures like insurance devices should be continuously revised and serve as a first measure for the empowerment of institutional development, the application of science, technology and knowledge, with monitoring and education-based targets. Appropriate governance and management of the multifaceted nature of risk are both fundamental if such considerations are to be factored into urban planning – and if existing risks are to be successfully mitigated, adapted or even transferred, as part of an integrated risk management approach employing preparedness, control and recovery. Reactive and proactive scenarios concerning possible hazards in urban areas should be clearly analyzed as opportunity-costs for society over the long term, especially with regard to global changes and regional impacts.

ACKNOWLEDGEMENTS

Gratitude is owed to book editors J.N. Parkinson, J. Goldenfum, and C. Tucci, and to Editors-in-Chief of UNESCO-IHP Urban Water Series Čedo Maksimović (Imperial College of London, UK), J. Alberto Tejada-Guibert (UNESCO) and Sarantuyaa Zandaryaa (UNESCO) for their useful suggestions. Discussions of insurance models are updated at the site of Integrated River Basin Group: www.shs.eesc. usp.br. This chapter is supported by Brazilian CNPq #310389/2007-1, #500159/2006-8 and FINEP-CT-Hidro #01.02.0086.00.

REFERENCES

Adger, N. 2000. Social and ecological resilience: are they related? *Progress in Human Geography*, Vol. 4, No. 3, pp. 347–64.

Andjelkovic, I. 2001. *Guidelines on Non-Structural Measures in Urban Flood Management*, IHP-V, Technical Documents in Hydrology No. 50, Paris: UNESCO.

Araujo, J.C., Doell, P., Krol, M., Guentner, A., Hauschild, M. and Mendiondo, E.M. 2004. Water scarcity under scenarios for global climate change and regional impacts in Northeastern Brazil, *Water International*, Vol. 29, No. 2, pp. 209–20.

Blaikie, P., Cannon, T., Davis, I. and Wisner, B. 1994. *At Risk: Natural Hazards, People, Vulnerability, and Disasters*. London and New York: Taylor & Francis Group.

Bull-Kamanga, L., Diagne, K., Lavell, A., Leon, E., Lerise, F., MacGregor, H., Maskrey, A., Meshack, M., Pelling, M., Reid, H., Satterthwaite, D., Songsore, J., Westgate, K. and Yitambe, A. 2003. From everyday hazards to disasters: the accumulation of risk in urban areas. *Proceedings of Workshop on Disasters, Urban Development and Risk Accumulation in Africa*, Nairobi, 8–10 January, UNDP-IIED.

Carmin, J.A. 2005. Disaster, vulnerability and resilience. Massachussetts Institute of Technology OpenCourseWare *Disaster, Vulnerability and Resilience*, Urban Studies and Planning, 2005, Lecture No. 1.

Costea, A.C. and Felicio, T. 2005. *Global and Regional Mechanisms of Disaster Risk Reduction and Relief: Review, Evaluation, Future Directions of Integration.* UNU-CRIS Panel Discussion on Competition and Complementarity Between Global and Regional Public Goods, United Nations, New York.

Crichton, D. 2008. Role of Insurance in Reducing Flood Risk. *The Geneva Papers* 33: 117–32.

Diamond, P.A. and Stiglitz, J.E. 1974. Increases in risk and in risk aversion. *Journal of Economic Theory*, Vol. 8, No. 3, pp. 337–60.

Galavoti, R., Mendiondo, E.M. and Vasconcelos, AF. 2007. Subsurface tree irrigation and carbon sequestration with urban wastewaters. *Proceedings International Conference Kalmar Eco-Tech 2007*, Kalmar, Sweden.

Goldenfum, J.A. and Tucci, C.E.M. (eds) 2005. *Proceedings Workshop of Urban Waters in Humid Tropics*, Foz de Iguacu, Brazil, 1–2 April. Porto Alegre, UNESCO/ABRH.

IFRC (International Federation of Red Cross and Red Crescent Societies). 2001. *World Disasters Report 2001: Focus on Recovery.* Geneva: IFRC.

IPCC (Intergovernmental Panel on Climate Change). 2001. *Impacts, Adaptation and Vulnerability, Report.* Working Group II, Summary for policymakers on Climate Change 2001, Geneva: IPCC.

ISDR (International Strategy for Disaster Reduction). 2002. *Living with Risk: A Global Review of Disaster Reduction Initiatives.* Geneva, (joint ISDR-Japan Gov-WMO-ADRC Publ.).

ISDR. 2005. Hyogo Declaration. *World Conference on Disaster Reduction*, 18–22 Jan, 2005, Kobe, Japan (Doc. A/CONF.206/6).

Jimenez, B. and Rose, J. 2009. *Urban Water Security: Managing Risks.* Paris: UNESCO and Taylor & Francis.

Lindell, M., Prater, C. and Perry, R.W. 2007. *Introduction to Emergency Management.* New York: John Wiley.

Mendiondo, E.M. 2005a. Scenarios of South American floods – From mitigating disasters to early-warning strategies. K. Takara et al. *Proceedings International Conference on Monitoring, Prediction and Mitigation of Water-Related Disasters* MPMD 2005, 12–15 Jan., 2005, Kyoto University, Kyoto. Kyoto: DPRI.

Mendiondo, E.M. 2005b. An overview on urban flood risk management. *Rev. Minerva Ciência e Tecnologia*, Vol. 2, No. 2, pp. 131–43. Sao Carlos, SP. Brazil.

Mendiondo, E.M. 2008. Challenging issues of urban biodiversity related to ecohydrology. *Braz J. Biol.* 68(4): 983–1002.

Mendiondo, E.M. and Valdes, J. 2002. Strategies for sustainable development in water resources systems. *Proceedings 2nd Int. Conf. New Trends in Water & Environ. Eng. Safety & Life*, Capri, Italy. Capri, Terra.

Mendiondo, E.M., Righetto, J. and Andrade, J. 2005. Mitigating floods through insurance. A.L. Aldana (coord.) *Proceedings International Course of Capacity on Monitoring and Forecasting of Hydrometeorological Processes*, Santo Domingo, Dom. Republic, Jul 2005. Madrid, CYTED-Prohimet.

Millennium Ecosystem Assessment, Scenario Working Group. 2005. Four scenarios. S.R. Carpenter, P.L. Pingali, E.M. Bennett and M.B. Zurek (eds) *Ecosystems and Human Wellbeing: Scenarios*. Vol. 2, Chapter 8. Washington: Island Press, pp. 223–94.

Simonovic, S. 2005. The disaster resilient city: a water management challenge. *Proceedings, Sustainable Water Management for Large Cities*. Simposium S2, VII IAHS Scientific Assembly at Foz de Iguacu, Brazil, IAHS Publ. 293, pp. 3–13.

Sivapalan, M., Takeuchi, K., Franks, S., Gupta, V., Karambiri, H., Lakshmi, V., Liang, X., McDonnell, J.J., Mendiondo, E., O'Connell, P., Oki, T., Pomeroy, J., Schertzer, D., Uhlenbrook, S. and Zehe, E. 2003. PUB 2003–2012: Shaping an exciting future for the hydrological sciences. *Hydrol. Sci. Journal*, Vol. 48, No. 6, pp. 857–880.

Tejada-Guibert, J.A. and Maksimović, Č. (eds) 2001. Frontiers in urban water management: Deadlock or hope?, *Symposium, Proceedings*. 18–20 June, Marseille, France. Paris, UNESCO. (Series IHP-V Technical Documents in Hydrology No. 45).

Thuo, S. 2001. The challenge of urban water management in Africa. J.A. Tejada-Guibert and Č. Maksimović (eds) *Frontiers in Urban Water Management: Deadlock or Hope? Symposium, Proceedings*. 18–20 June, Marseille, France. Paris, UNESCO. (Series IHP-V Tech.Doc. Hydrology No. 45) pp. 358–61.

Tucci, C.E.M. and Bertoni, J.C. 2003. *Urban Inundations in South America*. Porto Alegre, GWP/WMO/ABRH Public (in Portuguese).

UN (United Nations). 2006. Exploring key changes and developments in postdisaster settlement, shelter and housing, 1982–2006: Scoping study to inform the revision of *Shelter after Disaster: Guidelines for Assistance*, Paper UN OCHA/ESB/2006/6.

UNDP (United Nations Development Programme). 2004. *Reducing Disaster Risk: A Challenge for Development*. New York, Bureau for Crisis Prevention and Recovery.

UNEP FI (United Nations Environmental Programme Financing Initative). 2009. *The global state of sustainable insurance*. Geneva: UNEP FI.

UNESCO. 2007a. *Natural Disaster Preparedness and Education for Sustainable Development*. Bangkok: UNESCO Press.

UNESCO. 2007b. 6th workshop on urban water security, human health and disaster prevention. *Proceedings International Symposium of New Directions in Urban Water Management*. 11–14 Sept. Paris, UNESCO Headquarters.

WMO (World Meteorological Organization). 2006a. *Legal and Institutional Aspects of Integrated Flood Management: Associated Programme on Flood Management*. Geneva, Switzerland, WMO-No. 997.

WMO. 2006b. *Preventing and Mitigating Natural Disasters: Working Together for a Safer World*. Geneva, WMO Publ. No. 993.

WMO Bulletin. 2006c. *Preventing and Mitigating Natural Disasters: Working Together for a Safer World*. Geneva: Vol. 55, No. 1, January.

Chapter 7

Integrated urban water management: institutional, legal and socioeconomic issues

Arlindo Philippi Jr[1], Giuliano Marcon[2], Luis Eduardo Gregolin Grisotto[2] and Tadeu Fabrício Malheiros[3]

[1] School of Public Health, University of São Paulo, Brazil
[2] Centre of Information for Environmental Health, São Paulo, Brazil
[3] Engineering School of São Carlos, University of São Paulo, São Carlos, Brazil

7.1 BASIS FOR THE FORMULATION OF URBAN WATER MANAGEMENT POLICIES

Urban conflicts related to water use usually highlight problems of significant social, economic and environmental importance, particularly those relating to:

i) environmental health problems associated with pollution of water bodies and long-term degradation of strategic water basins,
ii) illegal occupation of environmentally vulnerable areas by communities where infrastructure and waste management systems are inadequate,
iii) ineffective water resource management systems, and
iv) lack of access to water of sufficient quality for different uses (domestic, commercial and industrial).

Due to the multidimensional nature and complexity of these conflicts, it is not possible to consider urban water management without a set of public and governmental policies that act in a coordinated way to support socioeconomic development and sustainable environmental management at local, regional and international scales.

Whilst scientific knowledge and technological advancements have improved the identification and characterization of various phenomena and processes related to urban water management, water resource policy and strategic plan formulation and implementation have not advanced at the same pace. However, considerable political and institutional efforts have been observed in various countries located in tropical climes, especially those in Latin America and the Caribbean.

Urban water management requires a process of articulation between different stakeholders who must act in an integrated way to harmonize the balance between the availability of water resources and the demand from domestic, commercial and industrial consumers. It focuses on the provision of adequate water quality conditions for the urban population and promotes sustainable use of water resources, taking into account all the corresponding social, economic and environmental interfaces within catchments. In essence, this means the inclusion of water resources within the institutional framework for urban environmental management.

Interrelationships in the urban environmental and political arena vary according to each water basins characteristics (physico-territorial, socio-cultural, environmental and institutional). Those factors, among many others, related to the demographic growth rate, thereby causing urban sprawl and slum expansion, are of particular importance. Additional factors are related to the impermeabilization of the urban environment and the resultant problems related to management of urban runoff and flooding. This phenomenon is common in large urban conglomerations and metropolitan regions in the humid tropics and becomes increasingly more complex to manage as conurbations grow.

There are three levels of analyses that depend primarily on the point of view of organizations responsible for setting related policies. Ultimately these determine the way in which water is managed in the urban environment.

1) *Urban water management:* involves a *locus* consisting of an urban area in which specific phenomena and conditions are enclosed and impacts occur in locally delimited areas. These areas are managed by local authorities who develop plans and programmes to implement management solutions and specific projects in response to and in support of local and national policies and strategic plans.

2) *Regional water management* includes a combination of urban and peri-urban catchments that require integrated policies, plans, and detailed guidelines for action that satisfy mutual interests. This is not simply a case of combining the various policies related to: (a) the use and occupation of the land, (b) environment protection, (c) water resource management, (d) environment sanitation, and other related issues. It necessitates a vision of the complex problems and demands that addresses the challenges related to the integrated management of urban water, encompassing one or more catchments.

3) *International urban water management* is relevant to issues that transcend hydrological boundaries, thus requiring the development of transboundary policies and guidelines for environmental protection. A number of important cases illustrate this situation in South America. For example, in the Prata River basin, both Brazilian and Paraguayan urban areas impact upon the hydraulic regime of the Paraná River, the downstream effects of which can be observed in Uruguay and Argentina. In the case of the Amazonian water basin, hydrologic and urban conditions in Bolivia and Peru affect the quality of downstream regions in Brazil (as in the case of the Madeira River in Rondonia and Amazon states). The Amazonian Cooperation Treaty was signed in 1978 (and is still in force), encompassing the many states that have 'border situations', especially in the Amazonian region. These states include the republics of Bolivia, Brazil, Colombia, Ecuador, Guyana, Peru, Suriname and Venezuela. In the specific case of transboundary basins, the need and strategic importance of urban wastewater treatment by means of bi or multi-lateral agreements requires negotiation between states and the harmonization of policies and legislation between the countries involved. Recent cases exemplify such a situation in South America: among them, the negotiation between Uruguay and Argentina in 2008 about cellulose and paper industry installations in the Prata river basin, and among Bolivia, Peru and Brazil on the conjoint analysis of impacts due to the implementation of two hydroelectric plants (Jirau and Santo Antonio) on the Madeira river with a total generating capacity of 6,450 MW.

Table 7.1 Evolution of demographic, and slum-dwelling population data in Latin America and the Caribbean

Population Data	Situation in countries of Latin America and the Caribbean	
	1990	*2001*
Total population (million)	440	527
Total urban population (million)	313	399
% Urban population/Total population	71.7	75.8
% Slum dwelling population/Total population	—	31.9
Slum dwelling urban population (million)	—	127

Source: Adapted from UN-HABITAT, 2003

The development of policies for urban water management is important for the establishment of strategies to solve complex themes inherent to urban waters. Spatial questions aside, attention is required to address the complex set of socio-political issues related to urban conurbations and metropolitan regions.

In this context, one of the main challenges influencing urban water policy is the issue of poverty, because poverty is intractable from illegal land occupation and the growth of informal, unregulated settlements. Thus, solving urban water problems is primarily an issue of finding adequate solutions to the massive prevailing problem of urban poverty.

This is an obstacle that has multi-dimensional constraints. According to the Brazilian Institute of Municipal Administration, 'Latin American cities have been built in a way that promotes profound social and economic inequalities, favouring the middle and high income neighborhoods and condemning the low-income population to a life or poor housing and inadequate services'. This is seen to be as a direct consequence of a lack of, or inefficient, public investment (IBAM, 2004).

According to data from the Economic Commission for Latin America (ECLA) (in OPAS, 1999), 39% of the households in Latin American countries live in poverty, 18% live in conditions of indigence, and 37% of dwellings were deemed uninhabitable. Many of these are found in metropolitan regions and conurbations. Examples are the peri-urban areas and slums in metropolitan regions in São Paulo and Rio de Janeiro (Brazil), Buenos Aires (Argentina), and Mexico City (Mexico), amongst many others. Data received from IBAM supported by reports from UN-HABITAT (2003) show that 127 million people lived in slums in 2001 (Table 7.1).

In the Brazilian case, approximately 63 million inhabitants live in slums located in metropolitan regions, representing approximately 46% of the total urban population. Many of these live in peripheral areas of urban conurbations in which population increases are between 3% and 6% per year. This condition is incompatible with the capacity of public authorities to develop and implement effective sector policies to respond to this reality (Bittencourt and Araújo, 2002).

Factors that contribute towards the increasing uncontrolled urbanization include the lack of capacity of government agencies responsible for public administration to invest in preparation of strategies, combined with the inherent difficulties of implementing urban development policies. These conditions favour the creation of irregular, low-income housing developments and slums that lack infrastructure for environmental sanitation, water supply and wastewater disposal.

The main characteristics of informal settlements in these urban peripheries are precariousness of housing, lack of infrastructure, degradation of the environment and the pollution of local water bodies. There are also many social problems related to low income, poor schooling, illegal activities and problems of public safety, amongst others.

Due to these socioeconomic complexities, it is not sufficient to promote purely technical solutions for urban water and wastewater management in densely urbanized slums and other poor informal settlements. There is an inherent need for integrated policies and strategies that are applicable to the complexities of problems in peri-urban areas. This is currently one of the main challenges facing local governments, managers and society in general.

Producing policies that formalize settlements and mitigate poverty is therefore vital for the implementation of water management policies in urban areas. These policies directly influence the design and implementation of plans, programmes and projects that make it possible to respond and attend to local demands and consequently promote improvements in the quality of the local environment.

These policies must also take into account solutions related to land ownership and the availability of credit for low-income households as a precondition for the success of projects to improve water and sanitation services.

7.2 ENVIRONMENTAL AND WATER RESOURCE MANAGEMENT IN THE URBAN ENVIRONMENT

7.2.1 Main progress and situation in the humid tropics

In recent years, areas of the humid tropics including the Latin America and Caribbean region, have watched while other parts of the world have made attempts to implement integrated water management systems. However, the majority of attempts have remained focused on sectoral use of water resources.

Until a few years ago, the vision of water resource management in South America was dominated by the belief that water-related environmental problems had yet to reach an intensity to justify a marked change in institutional framework and organizations – or the adoption of a different approach for the effective management of water resources (Bourlon and Berthon, 1998).

However, several recent experiences have reflected a change in this scenario. For the main economic and regulatory experiences and related instruments associated with water resource management in Latin America and the Caribbean, the work of Seroa da Motta et al. (1996) is considered to be the most important. The authors draw upon the work of Huber and colleagues from the World Bank in the Latin America and Caribbean Environment and Urban Development Division.

Huber analysed the situation in ten countries: Barbados, Brazil, Chile, Colombia, Ecuador, Jamaica, Mexico, Peru, Trinidad and Tobago and Venezuela. In these countries, institutional fragility was clearly identified as the main barrier to the development of effective management arrangements, which would engage with society in discussions determining the strategic decisions of local impact, and the budgetary and fiscal implications of these decisions.

Out of the analysed countries, Barbados and Ecuador lacked any form of national environmental legislation; Barbados, Jamaica and Trinidad and Tobago lacked a government ministry responsible for environmental management; Barbados, Ecuador, Jamaica, Peru and Trinidad and Tobago lacked a specific chapter on the environment in their Constitutions. And finally, Barbados, Ecuador and Peru did not have any specific executive agency to enforce environmental legislation.

According to Seroa da Motta et al. (1996), 'the water and sanitation companies are usually controlled by the state government, whilst the collection of solid waste is administered mainly by the municipal authorities'. Countries such as Brazil, Venezuela, Colombia and Mexico, on the other hand, have more consolidated and decentralized structures; many of which have communication channels for inter-sectoral articulation.

Given growing political and social pressures, services and demands related to the environmental sector have seen accentuated growth without the accompanying necessary fiscal support for their operation and maintenance. This fact exacerbates problems in defining management goals, standards and instruments, which are consequently established above the capacities of the public agencies for administration, environmental monitoring and regulation.

As a consequence, in Latin America and the Caribbean, urban management systems are traditionally supported by dated, inadequate standards, regulations and laws to limit pollution and ineffective control actions, which impact adversely upon the quantity or quality of natural water resources.

7.2.2 Financial considerations

The main financial instruments that have a direct application on urban water management are the following:

- tax and credit-related incentives
- cost-recovery fees
- charging for use of the water, and
- marketable licenses.

The tax and credit-related incentives in countries such as Brazil, Mexico and also in Colombia (see Box 7.1) include subsidies towards credit for investments in industrial pollution control. In Colombia and Brazil, there are reductions and exemptions on income taxes for the adoption of clean technologies. These types of financial incentives are also seen in the case of Venezuela and Jamaica.

Cost recovery corresponds to increases in water and sewerage fees, potentially with the use of cross subsidies, in order to expand service coverage. Chile and Colombia have adopted this approach on the basis of long-term marginal costs. This is a subject that assumes strategic importance when considering the involvement of the private sector in the management of sewerage services.

For charges relating to excessive use of water, some peculiarities are observed in Mexico and Colombia, which both have effective tariff structures in force. In Mexico, the imposition of fines for pollution has been carried out since 1991 in accordance with legislation that makes it possible for the National Water Commission (NWC) to apply the 'polluter pays' principle to users (including municipalities, industries, etc.)

Box 7.1 Charging for effluent discharge in Colombia

Since 1974, Colombia adopted a charging system based on water usage and a separate system for effluent discharges. Collection is coordinated through regional agencies, but this presents problems related to the lack of appropriate information on impacts, and incompatibility with the data collection system with the planning systems.

These obstacles have subsequently weakened the collection process, resulting in an accumulated total of US$116,000 out of a potential income of US$90 million. However, despite all the problems inherent to the process, in the few cases where difficulties were overcome, changes in the patterns of water use and reductions in consumption and pollution were observed.

Currently, with a new approved legislation, Colombia has modernized its collection system, adopting the principle of 'total environmental costs' where the amount of the tariff varies in accordance with the associated environmental services and the cost of damage to the environment. Implementation of this system, however, requires the need for a more solid institutional apparatus, especially to define the values of the tariffs.

---

that exceed standards of organic matter or solid particles in suspension. But, for discharges below 3,000 m³ a collection system based on volume and not on pollutant loads is adopted.

However, this system may fail in its application as a result of deficient monitoring systems, accompanied by a general lack of reliable information or careful analysis of possible impacts. According to Seroa da Motta et al. (1996), the national coverage of the water system exerts too many resource and financial demands on NWC to introduce an effective system for monitoring. In addition the system is compromised by a lack of enforcement with regard to entities who refuse to pay fines. Furthermore, this is hindered by a lack of public awareness and motivation to participate in legal processes against those responsible for the pollution.

Marketable licenses are instruments that exist only in Chile, although Peru has discussed their adoption. These are similar to the granting of user rights, but with the difference that they are marketable. The Chilean experience with marketable licenses of water use dates back to the 1920s, but its legal base was only established within the Water Code of 1951. This allowed the State to grant concessions to private entities according to water use-related priorities. Water transfers were allowed, if the water use remained the same. Later, in 1969, commercialization of concessions was prohibited, as water was considered a state property. But the Water Code of 1981 reintroduced permanent rights on water and, with the view that they were separate from rights on the land, asserted that they could be marketed for any use.

7.2.3 Legal considerations

Currently, more than 300,000 users of water resources is enrolled in Chile, but only between 35% to 50% have legal ownership. However, in the light of the Chilean tradition with regards to low commercialization of users' rights, expectations are strong that the system will demonstrate a certain stability, and the process a greater credibility, in which the capacity to negotiate the rights of water use will exist even to non-ownership holders.

In short, the review of the international experience has a key meaning for what Gutierrez (1998) calls 'environmental harmonization', in which the coordination of policies and instruments that aim to reduce differences among countries are central to the introduction of changes at international level.

In Brazil, the development of the water resources management sector was supported historically by the use of fluvial potential to generate hydroelectric power. The centralized legal and institutional apparatus was characterized by the principle of rationalizing use and the protection of natural resources in accordance with the needs of industrial expansion. To this effect, water and electric power were deemed as subjects of national security.

Brazil has consolidated the formulation of principles for the multiple use of water resources through its Water Code (1934). The approval of the Water Code was followed by the creation of countless agencies intended for water administration, but predominantly for hydroelectric purposes (Barth, 2002). The theme of urban sanitation began to incorporate an environmental dimension in relation to water resource protection in the 1970s with the elaboration of the National Sanitation Plan (PLANASA).

The most recent Federal Constitution of Brazil (1988) and the creation of the Water Resource General Office in the Environment Ministry in 1994 were forerunners of federal environmental policies for water resources. These were finally institutionalized in 1997 by Federal Law no. 9.433 (ABRH, 1997), which represented an important institutional milestone in the country for planning and administering water resources, and as a prerogative for sustainability.

Amongst the basic principles described above, of key importance are the principle of the economic value of water, which promotes a more rational use of water; as well as other aspects related to the multiple use of water, such as the need for decentralized and participative management, and the adoption of the watershed as a planning and management unit.

In relation to these principles, the following are identified as instruments in Brazilian policy for transboundary water resource management:

- water resource planning
- classification of water collections according to their specific uses
- grants for the use of water resources
- tariff collection according to the actual use of water resources, and
- the development of information systems on water resources.

In addition to the Brazilian cases referred to above, several situations in other countries in the humid tropics exemplify cases where solutions to conflicts over the use of water resources have been seen to have a direct impact on urban water users. These are described in the following section.

7.2.4 Actions and initiatives in Latin America and the Caribbean

In Latin America, several experiences deserve equal prominence, including the projects of 'Support to the Improvement of Quality of life in Irregular Settlements' of Chile,

'Urbanized Lots for Low Income Population of the North Area' of Argentina, and 'Union of Tenants and Housing Applicants, Veracruz (UCISV-VER) Housing Programme for the Peripheral Areas of Xalapa, Veracruz' in Mexico (IBAM, 2004). A significant focus of these programmes is the improvement in the infrastructure and services in poor urban areas. These are related to the goals and principles of the 'Urban Management Programme for Latin America and Caribbean', co-coordinated by UN-HABITAT and the United Nations Development Programme. There is also the UNESCO International Hydrological Programme (UNESCO-IHP) for Latin America and the Caribbean, which aims to improve the information available on water resources, and to increase knowledge on the world water cycle, favouring more sustainable development and management of water resources.

In addition to structural actions, other initiatives have been introduced in Latin America relating to the management of urban waters. Mediondo (2004) mentions examples of water-solidarity projects, documented and disseminated by the 'Integrated Centers of Watersheds', such as the Centre of Information for Environmental Health (NIBH) of the Engineering School of São Carlos of the University of São Paulo (EESC-USP), Brazil, which supports integrated, interdisciplinary and inter-institutional teaching, research and extension actions through case studies and 'watershed-schools'. These are courses for students, researchers, practitioners, government representatives and other stakeholders which aim to integrate different research disciplines and promote working partnerships and team collaboration. This involves precautionary practices as part of a planning exercise in which solutions are discussed in relation to ethical association, multi-cultural spheres and plurality of visions around water resources.

However, in general the management of urban waters in the humid tropics lacks the larger scope of environmental policies and water resources, with few plans, programmes and projects adopting a truly integrated approach that address the complexity of territorial occupation in urban centres of these countries.

7.2.5 Proposals and actions developed in Brazil

The analysis of the policies and systems in Latin America for management of the environment and water resources in countries in the humid tropics makes evident the need for specific treatment on the subject of urban waters. This is particularly the case with regard to the difficulties and relative complexity of defining and introducing feasible integrated management strategies.

Some of the main problems within this theme are the political-institutional, administrative and economic limitations for developing plans, programmes and projects. Initiatives that are appropriate for mega-cities and metropolitan regions must include poverty reduction as a key factor for attaining the intended sustainability. Another important restriction relates to the financing of interventions in urban areas, considering that the nature of these actions is long-term and multi-institutional, and necessitates significant intersectoral involvement.

To make it easier to overcome these obstacles, most of the programmes and projects are linked to the credit of multilateral financing agencies such as the International Bank of Reconstruction and Development International (IBRD), the Japan Bank for International Cooperation (JBIC) and the Inter-American Development Bank (IADB).

In Brazil, programmes such as the Basic Sanitation Programme for Low Income Populations (PROSANEAR), the Social Action Programme, the Water Quality Programmes (PQAs) and HABITAR BRASIL, amongst others, contribute towards increasing the flow of resources and investments. These actions are characterized by urban restructuring, including the structural modification of public and private spaces, the construction of housing and infrastructure development.

Consequently, for Brazil, programmes and projects considered to be pioneering in the promotion of an integrated approach have been partly financed by multilateral organizations. These include the Programme of Environmental Sanitation of the Watershed of Guarapiranga, in the Metropolitan Region of São Paulo, Brazil; Slum-Neighbourhood Project, in the city of Rio de Janeiro, and also the following:

- *Modernization Programme of the Sanitation Sector* (PMSS): an instrument of the National Sanitation Policy, intended for institutional modernization, extension and improvement of the systems and bases of information on environmental sanitation and water resources.
- *Social Action Programme in Sanitation* (PASS): financed by the Inter-American Development Bank (IADB) and aimed at increasing coverage indexes and the quality of services related to environmental sanitation rendered by companies and concessionaires.

The financing of integrated programmes by multilateral institutions provides one way of tackling administrative inertia and the lack of urban institutional motivation, promoting reorganization and the cooperation between institutions at different decision-making levels around common interests. This, however, is not a panacea, as there are several aspects to be overcome, among them:

- The incapacity of public authorities to organize and manage technical teams, systems and databases, financial resources, and their inefficient administrative and organizational apparatus for preparing programmes.
- The difficulties and legal impediments related to the introduction of intersectoral initiatives, considering that thematic disciplines related to the management of urban settlements (such as housing, urbanization planning, environmental sanitation, and water resource management) are usually linked by various regulatory instruments and procedures. This situation can lead to problems related to the coordination required for the successful implementation of projects. The complexities related to the analysis and approval of integrated programmes are often exaggerated when there are various multilateral organizations involved.

The diverse political and inter-institutional relationships that are characteristic of Latin-American countries are strongly influenced by the overall political and macro-economic context. Some projects coordinated by the Brazilian National Water Agency (ANA) aim to promote the integration of efforts, but are still restricted to water resources management, mainly in urban areas. This is the case of the Basin Restoration Programme (PRODES) whose objective is to:

- reduce the critical levels of water pollution from domestic, commercial and industrial sources in highly urbanized catchments

- implement new management systems for water resources in these areas, through the constitution of Watershed Committees, and
- encourage respective agencies to develop and implement mechanisms to define the right to water resources, in accordance with the National Policy of Water Resources.

Through PRODES, ANA offers subsidies to wastewater treatment plant operators in heavily polluted catchments who can demonstrate that they have achieved agreed targets for pollution reduction.

It is also important to highlight the importance of the 'polluter pays' principle as a mechanism for achieving improvements to the quality of nature water resources through the application of cleaner technologies. It may be applied to systems of wastewater management and environmental sanitation, including pollution control infrastructure in its different forms, drainage of surface waters and retention of floods (Carrera-Fernandez and Garrido, 2002). In this case, a portion of the financial requirements for adapting the management of urban waters would be supplied by this instrument.

Other actions supported by ANA are also prominent, including the 'Programme of Sustainable Urban Drainage' of the Federal Government, which aims to produce updated information on the management of urban waters. This activity is coordinated by the National General Office of Environmental Sanitation (SNSA) and forms part of the cooperation activities between Brazil and Italy, with regard to the dissemination of new concepts for the management of urban waters.

Table 7.2 presents a summary of the problems and causes associated with the subject of urban waters according to discussions found in the text above.

Table 7.2 **Some key problematic issues in urban water management and their causes**

Problems	Causes
Low priority attributed to the subject of water resources	Lack of political interest and competing sectors demanding use of limited financial resources
Poor coverage and access to infrastructure and services	Rapid urbanization, social exclusion, lack of institutional capacity to plan for infrastructure expansion, insufficient investment capacity in relation to demand
Poor operation and maintenance of existing infrastructure	Lack of effective cost-recovery mechanism and willingness to pay
Poor skills and technical capacity	Lack of training and poor salaries
Increasing environmental emergencies (particularly flooding)	Climate change, lack of monitoring, and poor system for flood preparedness
Contamination of water supply systems and bodies	Lack of good sanitation and effluent treatment
Increasing environmental emergencies with high incidence of urban inundations	Absence of a strategic approach, which would enable the system to, through a process of continuous evaluation, incorporate successful experiences and reorient efforts in a synergic way

7.3. URBAN WATER MANAGEMENT IN THE HUMID TROPICS: A NEW PARADIGM

7.3.1 Institutional reforms

The idea of a new approach to the management of urban waters has, as its starting point, the paradigm that water is a finite natural resource, with social and economic value. Adopting this approach helps to identify efforts that are necessary to reverse actions contributing to the lack of sustainability in the context of urban waters and their interface with public health, quality of life and environmental protection.

One of the recommended actions to achieve this objective is the strengthening of inter-institutional relationships between countries, promoting the harmonization of legal frameworks, with special attention to the effective and efficient management process of urban waters towards achieving environmental sustainability.

Common legal structures for defining general principles and guidelines can also be established for defining specific public policies related to urban waters, and to provide solutions to conflicts related to urban water management in Latin-American countries.

This synergy of efforts and responsibilities assures a more favourable political–institutional environment for attracting public and private investments, international grants and loans, making credible the implementation of management policies for urban water. However, it is the combination of national contexts and the specifics of local environments that present significant challenges related to the development of integrated and strategic policies for managing urban waters.

Many of these challenges are greater in unplanned areas where low-income populations have settled. In Brazil, given the specifics of slums and peri-urban areas (e.g. lack of planned access routes), the costs for implementing integrated solutions tend to be higher than in more consolidated urban areas.

Overcoming this problem requires greater efforts between urban planners, engineers, social scientists and community facilitation specialists. It is essential that projects applicable to urban waters, such as the implementation of sewerage networks, water supply systems, collection and disposal of solid waste, drainage, and treatment units develop in a way that is compatible and integrated with other projects designed to engineer the environment and bring about improvements to other forms of urban infrastructure and services.

7.3.2 Political commitment and policy development

This integration, from the perspective of public administrations, requires the structuring of policies and actions supported by proper institutional arrangements, which assure the allocation of responsibilities to the appropriate agencies involved. There is therefore a need for institutional training support to ensure sustainability of those policies and the corresponding actions. As described below, this also requires the participation of local stakeholders and communities (potential beneficiaries) in the planning and implementation of those strategies.

At the regulatory level, the development of appropriate instruments such as levels of service standards, discharge permits and licenses to operate is required. However, in advance of this, there is a need for higher-level political acceptance and authorization

---

Box 7.2 Santa Cruz + 10 Declaration

The Santa Cruz +10 Declaration links the objectives of sustainable development with the pro-motion of good governance, institutional capacity strengthening and the formulation of national legislation related to environmental management for their effective application. The declaration agreed to the following resolutions:

(i)   Extend access to drinking water and sanitation services to all people in the jurisdiction of each Member State based on environmental non-discrimination, solidarity and sustainability.

(ii) Promote, as appropriate and with the consent of the participant states, the accomplish-ment of studies, plans, projects and combined actions for sustainable protection and use of surface and underground water resources, of the ecosystems of humid zones and the related biodiversity.

(iii) Promote the integrated management of water resources via management arrangements based on public participation, institutional transparency and access to environmental information.

---

for the need for institutional reform, which requires policy statements to support the reform process.

At the national level the reform process may often be stimulated by an agreed inter-national commitment, such as Agenda 21 (Chapter 18, Item 18.50), which presents guidelines to ensure that states implement activities to improve water resources, allot-ting 'more attention to the badly assisted rural zones and to the low-income urban suburbs' in accordance with their capacity and available resources and through bilat-eral or multilateral cooperation.

Another example that focuses specifically on the Latin American context, was the Santa Cruz +10 Declaration, which was conceived in December, 2006, at the First Inter-American Meeting of Ministers and High Authorities of Sustainable Development of Latin America in Santa Cruz de La Sierra, in Bolivia. The declaration is described above in Box 7.2.

7.3.3   Participatory processes in support of the reform process

The insertion of transparency is an important component of the reform process asso-ciated with the previous actions described above. This requires the engagement of key stakeholders from government, the private sector and civil society as part of the process of reversing established problems, and in the development of strategies to implement new plans.

Engaging and enabling local stakeholders to become actors in the implementation process requires the integration of policies, plans, programmes and projects that con-tribute within a systematic approach to improvements to the local environment, and the instigation of better quality water and sanitation services. Water is an issue that can bring together stakeholders from different sectors to identify priorities of mutual benefit, connect policies, and develop the strategic and participatory planning tools to help negotiate and resolve conflicts

Once participation is constituted as a fundamental element in the sectoral systems of environmental management relating to water and urban resources, it is possible and advisable that actors state in an effective way their common interests on behalf of integrated actions that avoid sectoral isolation. Cases such as the Water War, which took place in 2004 in Bolivia, or the World Panel on Financing of Water Infrastructure, formed in 2001 by Global Water Partnership (GWP), the World Water Council (WWC) and the World Water Forum (WWF) are references that the future of urban water management may use to base participatory approaches with a particular emphasis on the role of women in the process.

7.3.4 The role of information

When considering the complexity of the subjects involved in the context of urban water management, and its relationship with social, economic and environmental components, it is mandatory to establish and structure information systems and a set of indicators to support the formulation, implementation and evaluation process of urban water policies.

These systems should favour the integration of information from the various sectors related to the subject of urban water, understanding that the effectiveness and sustainability of management depends on the appropriate operation of the components. These include water supply and sewerage services, drainage and solid waste disposal systems, territorial planning, the water resource system and transport systems, amongst others.

A suitable information system, with integrated indicators, favours understanding of the way in which the different components work and what potential risks they can pose to the management of urban waters. The appropriate use of indicators is a key factor because it permits the creation of a connection or potential causal relationships amongst degrading activities or inadequate management of the components of environmental sanitation. This enables a better understanding of the impacts on water supply and sewerage systems, creating favourable conditions for the placement of efforts to reverse the cause of these problems.

These same external pressure indicators on water supply and sewerage systems make possible better sectoral strategic planning, responding to the rising demands of society. The appropriate use of interface indicators for both water and sewerage services enables benefits that act in two ways: promoting more effective policies and efficient actions in urban planning, and active cooperation in the management of water and sewerage services.

This integration presupposes the use of easily available information in each of the sectors. But in reality, only part of this information is likely to exist, and in most instances there will be a need to improve the information base.

7.4 CONCLUDING REMARKS

The subject of urban water is an obligatory theme in the agenda of all governments that have a commitment to promoting quality of life and public health. To achieve this goal requires promotion of the practice of integration for all main actors in the urban arena. The analysis of policies for water resource management and their respective technical, economical, legal and institutional management instruments highlights the need for much greater efforts and resources to support intersectoral collaboration.

It has been observed that the formulation and implementation of policies and strategies related to the management of urban water must be based upon an integrated understanding of the different environmental components, considering the capacity of the ecosystem to provide resources and assimilate wastes.

Urban poverty is one of the main challenges for solving water-related problems and conflicts in Latin American countries and other developing countries in the humid tropics in other regions. The formulation of effective policies requires much more than a consideration of technical issues. Socioeconomic vulnerability is a vital component and comprises aspects relating to poverty, social exclusion and the capacity of the society to organize itself.

REFERENCES

ABRH (Associação Brasileira de Recursos Hídricos). 1997. *Política e Sistema Nacional de Gerenciamento de Recursos Hídricos*. São Paulo: ABRH.

Barth, F.T. 2002. O Modelo de Gestão de Recursos Hídricos no Estado de São Paulo. A.C. de M Thame (ed.) *Comitês de Bacias Hidrográficas: uma Revolução Conceitual*. São Paulo: IQUAL Editora, pp. 17–30.

Bittencourt, A. and Araújo, R.G. 2002. *Avaliação do Setor de Abastecimento de Água e Esgotamento Sanitário no Brasil: Os Problemas do Atendimento às Populações Urbanas Pobres e do Controle da Poluição Hídrica. A Agenda Ambiental Marrom*. São Paulo: Banco Mundial.

Bourlon, N. and Berthon, D. 1998. Desenvolvimento sustentável e gerenciamento das bacias hidrográficas na América Latina. *Água em Revista: Revista Técnica e Informativa da CPRM*, Vol. VI, No. 10, pp. 16–22.

Carrera-Fernandez, J. and Garrido, R.J. 2002. *Economia dos recursos hídricos*. Salvador: Edufba.

Gutierrez, M.B.S. 1998. *Desenvolvimento Sustentável no Mercosul: a Proposta de um Marco Regulatório*. Brasília: IPEA – Instituto de Pesquisa Econômica Aplicada.

IBAM (Instituto Brasileiro de Administração Municipal). 2004. Urbanização de Assentamentos Informais e Regularização Fundiária na América Latina. *Foro Iberoamericano e do Caribe sobre Melhores Práticas*.

Mediondo, E.M. 2004. *Gestão Hídrica Sustentável em Bacias Sulamericanas para o século XXI: Desafios da Hidro-solidariedade em Projetos Transnacionais*. Relatório do NIBH-SHS/ EESC/USP. São Carlos.

OPAS (Organización Panamericana de la Salud). 1999. *Atenção Primária Ambiental- APA*. Brasília, OPAS/BRA/HEP.

Seroa da Motta, R., Ruitenbeek, J. and Huber, R. 1996. *Uso de instrumentos econômicos na gestão ambiental da América Latina e Caribe: Lições e Recomendações*. Brasília: IPEA – Instituto de Pesquisa Econômica Aplicada.

UN-HABITAT (United Nations Human Settlements Programme). 2003. *Slums of the World: the face of urban poverty in the new millennium*. Geneva: UN-HABITAT.

Chapter 8

Education and capacity-building

Jose Ochoa-Iturbe[1] and Jonathan Neil Parkinson[2]

[1]*Director-School of Civil Engineering, Universidad Catolica Andres Bello, Caracas, Venezuela*
[2]*International Water Association, London, United Kingdom*

8.1 INTRODUCTION

8.1.1 Changing conditions in the urban water cycle

In recent years, an unusual number of extreme events have occurred throughout the world, either due to anthropogenic activities, climate change or a combination of causes. Although widespread media coverage and internet use may create the impression that such events are more numerous than in actuality, it is also increasingly apparent that the frequency of occurrence and the magnitudes of flows have actually increased significantly.

For example, although flooding is highly variable, records taken since 1950 show an increase in the occurrence of disasters. Figure 8.1 clearly indicates an increase in major natural catastrophes since 1950, leading to considerable loss of human life (50,000 deaths in 1999) as well as significant overall economic losses (WWAP, 2003). The United Nations World Water Development Report (2003) states that about 65% of people affected by natural disasters were affected by flooding. This is considerably more than any other natural event, including famine (20%), although it is acknowledged that famine is responsible for a higher number of fatalities.

The extent of damage has also increased, in terms of damage to infrastructure as well as loss of human life. In addition to intensive rainfall, risks and impacts are exacerbated by the fact that houses are constructed – often informally by poor families – on flood-prone areas or unstable slopes which, deprived of vegetative cover, are susceptible to landslides in times of heavy rainfall.

More attention is now being paid to the prediction of flood events and the corresponding design of infrastructure to protect against their effects and diminish their impacts. But, according to UNESCO's Division of Basic and Engineering Sciences, for every US$100 spent by the international community on risk and disasters, US$96 go towards emergency, relief and reconstruction, while only US$4 is spent on prevention. Yet, each dollar invested in flood prevention reduces by up to US$25 the losses incurred in the case of natural disasters. Therefore, from a purely economic perspective, investment in prevention should form part of national government spending.

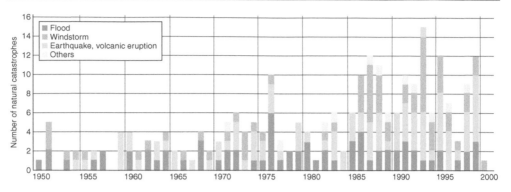

Figure 8.1 Trends in major natural catastrophes, 1950–2000

Source: Munich Re, 2001

A crucial part of the prevention process involves educating people about the water cycle and the different aspects of water management. It is also necessary for engineers to be adequately skilled in the design and construction of flood protection infrastructure, and also to liaise with local stakeholders – particularly those at risk from flooding – so as involve them in decision processes concerning flood mitigation and response strategies.

A better understanding of the water cycle is essential for ensuring the constant and consistent satisfaction of water demands – to adequately manage the amount of available water in our regions, especially during times of flood and drought, and to adequately design works to cope with extreme events. In tropical climates, weather conditions may change very rapidly, a fact that has to be considered in particular in areas prone to damaging flash floods.

This chapter is dedicated to the human resources requirements necessary to confront and manage the widespread problems related to urban drainage occurring in most large urban centres in the humid tropics. It focuses mainly on experiences from the Americas, but is also relevant for situations in other regions of the world.

8.1.2 Assessment of needs

Large-scale flooding is usually the result of extreme rainfall events, but also sometimes occurs as a result of continuous low-intensity rainfall, falling on saturated soils over several days and followed by a heavy storm. In certain areas, non-meteorological floods may also be triggered by rapid snowmelt and/or high seismic or volcanic activity.

Since extreme events are subject to chance occurrence, engineers can only predict probabilities of magnitudes and frequency, using statistical tools and based on actual data. The greater the available data, the higher the probability of accurate results. As Sir Charles Pereira states in an introduction to a book by Gunston (1998): 'The indispensable key to water resources management is measurement'.

The accumulatation of data of sufficient quantity and quality for the application of numerical models is therefore the first requirement for understanding these phenomena. A prerequisite is the existence of a hydrologic network (or its densification if such a network already exists), particularly in catchments at greatest risk of flooding. Gauging stations should be located according to World Meteorological Organization guidelines (WMO, 1996) to ensure complete coverage of the basin. These stations should include instruments for taking rainfall and stream flow measurements, and so on. The installation of equipment for networks is time-consuming work and should commence as soon as possible. As discussed below, the personnel needed to operate these stations also need training, which can also be a time-consuming exercise.

Flood zoning is an integral component of urban water management. The definition of flood zones necessitates up-to-date maps which accurately reflect topography and land use. However, in developing countries (with few exceptions), much of the data is scarce, dated and generally of poor quality. This acts as a constraint on the identification of risk zones and hydrological modelling. Remote-sensing technologies combined with Geographic Information Systems (GIS) can be used as a tool in this process. But this may be overly ambitious without additional investment in capacity strengthening of organizations and further staff training using external expertise.

More extensive datasets, describing in detail rainfall distribution, runoff regimes (land occupation, soil types, etc), and other related hydrological topics, lead to greater accuracy with regard to the prediction of events and the magnitude and frequency of occurrence. This in turn results in more effective flood mitigation strategies.

In the light of the above summary of key physical elements required for the study of drainage problems, especially in urban areas subject to flooding (and heavy losses), we now focus in more detail on the training and education of water professionals who are required to respond to and solve these problems.

8.2 FORMAL EDUCATION

8.2.1 School education

Attempts have been made to introduce concepts of environmental awareness in schools. However, this trend is relatively recent in most, but not all, countries, and therefore many children, especially in communities where school attendance is low, are still growing up with little awareness that water is a limited resource. Even countries who are ahead of the curve began the education process, at most, only two decades ago.

Therefore, there remains a real need to educate children on the value of water and to raise awareness of the importance of conserving this resource through sustainable practices. School programmes from nursery to high school need to place greater emphasis on the water cycle and, in particular, precipitation and runoff, as these two components are perhaps the most important to human development.

One example from North Dakota in the United States established initially in 1984 has now been expanded to seventeen countries. The Water Education for Teachers (WET) project has developed an extensive array of material (available for download from www.projectwet.org) and seminars to help teachers communicate to children the real value of water management.

8.2.2 Higher education

In recent years, there has been a decline in the number of civil engineers worldwide. This is partly due to the migration of students to newer, potentially more exciting disciplines, such as those involving information and communications technologies, which attract higher salaries. This has left the profession poorly equipped to deal with the increasing challenges it faces. This situation is particularly evident in developing countries, where large infrastructure projects are usually undertaken by foreign professionals due to the lack of a qualified national workforce.

Of particular concern is the fact that Bachelor degree courses often leave out the topic of urban hydrology and drainage in favour of other topics which are considered to be of greater interest, and thus more attractive to prospective students. Although the basics of fluid mechanics or hydraulics should be sufficient for handling most drainage problems, the state of the art depends very much on empirics, hands-on experience and expertise. Therefore instructors in this discipline should be experienced professionals and practitioners with a good understanding of the important linkages with other disciplines key to integrated urban water management.

As mentioned above, drainage is generally incorporated within hydraulics courses (if included at all), therefore students receive little specific instruction on the topic. Some engineering schools have the subject as an option in their final semester, but only a limited number of students are committed enough to study the subject in sufficient detail to then move on to become practitioners. Urban water management needs to be more intrinsically encapsulated within the curricula, particularly as flooding is becoming more extensive and the impacts of climate change are becoming increasingly felt. As a result of this, most civil engineers will be involved in the design of some form of drainage or flood defense infrastructure at some stage in their professional practice.

In order to design the necessary works, engineers require extensive datasets describing rainfall, surface types and dimensions/slopes of drainage channels. These enable them to study the phenomena, carry out modelling, and undertake risk analysis. Therefore, civil engineering curricula, especially in the area of fluid mechanics and hydraulics, needs to emphasize drainage problems, in addition to more traditional programmes, including lectures on risk analysis and participatory planning, which are essential, and important topics requiring collaboration with different disciplines.

Engineering education is also undergoing a transformation due to advances in computer technologies and software, and the use of GIS and the internet. New approaches towards educational methods in the classroom or for distance learning are rapidly emerging and are replacing traditional teaching methods. However, a review of civil engineering courses in the United States showed little change over the last fifty years in terms of overall content and structure (Ligget and Ettema, 2001). Morever, exposure to the wide variety of available software may be limited by the instructor's personal experience.

8.3 OTHER TRAINING NEEDS

8.3.1 Training for other water management technicians and professionals

This section focuses on a discussion of the training needs of staff other than the engineers working for organizations and agencies involved in urban water management. These groups comprise environmental managers, regulators and economists, as well as civil servants. These non-engineering professionals need to understand basic water management principles and acquire certain concepts from experienced professionals in the subject. However, instruction should not be overly technical and should be performed in accordance with their line of work.

Senior managers from water utilities and agencies with overall responsibility for the management of different components of the water cycle within a catchment (including water supply systems and wastewater treatment plants, as well as sewerage and drainage infrastructure) need to understand the seasonal changes occurring in water quantities and the impact on the timing of occurrence and frequencies of extreme events, as well as other relevant information in order to properly manage the resource.

Administrators from these organizations are seldom drawn from engineering backgrounds. As such, they need to have a better understanding of the essentials of water management from a technical perspective in order to equip them to make better decisions. For example, decisions related to evacuation from high-risk zones are difficult if the magnitude of the event or the potential consequences are unknown. Training in the urban water cycle and drainage basins, including principles of extreme events, such as probability and risk analysis, should be offered to these staff. This will assist with the design of urban storm master plans, including land-use zoning, which forms an important part of flood mitigation. Other staff from utilities such as operators responsible for handling equipment like pumps, or for measuring different parameters (especially hydrologic data), need to understand the fundamental hydraulic principles involved in their work to be able to better communicate with professionals doing the actual calculations and design.

8.3.2 Civil society and NGOs

Members of civil society, community leaders, NGO staff and the general public have a more practical need, which deals with the effects on society caused by extreme events. Education of the general public on floods, how they are generated, and potential risks and damages is therefore of particular importance.

The level of detailed information required varies according to the education of those involved. The majority of the general public has limited need or capacity to assimilate detailed information. But, for the mobilization of residents in flood response strategies, they need to be better able to interpret flood maps and identify flood-prone areas, as well as be aware of alarm systems, evacuation routes and safety zones.

The poorer sectors of society are usually the worst hit by catastrophes, partly because of the land they occupy, and partly because in many cases little information is given to them on how to respond. What makes it worse is that the poor are also the least capable of recovering economically from flood impacts. Simple early warning systems, such as water marks showing levels of flooding from past events, will therefore help to raise awareness, and could help the evacuation of such people in flood-prone zones.

8.4 INFORMATION AND DATA REQUIREMENTS

8.4.1 Information requirements for engineers

Urban drainage systems are an integral part of the urban water cycle, which has to be monitored and analysed under different scenarios in order to be better understood and managed. To do this requires continual measurement of rainfall, and flows in creeks, rivers and other water bodies so as to record peak flows and times of occurrence. Once this is done, an engineer is needed to analyse the data prior to the planning and design of interventions.

Drainage engineering is more or less defined in textbooks handbooks and guidelines, with regards to the calculations for dimensioning and construction of storm runoff facilities. The existence of many software packages that perform storm drainage calculations facilitates the design of storm drainage structures using standard methods of calculation. All need basic input data describing topography, rainfall patterns and details of drainage conduits (cross-sectional area, slope and roughness coefficients, etc). More sophisticated models require additional parameters for soil type, infiltration and evaporation rates, and so on.

However, it is important to note that the use of models by inexperienced personnel who do not appreciate their limitations and restrictions is potentially problematic. For example, in urban surroundings, flows of runoff during extreme storm events do not follow the drainage pathways associated with constructed drainage infrastructure. The overland flow of floodwater adds a new dimension of complexity for those trying to analyse and understand local urban hydrology.

In addition, if a river or creek runs through a city, which is a common occurrence, studies have to be made of the whole basin with emphasis on peak flows, time to peak and flood-prone areas, as well as the localized area which forms the focus of attention. This should avoid the design of infrastructure restricted to a specific area which does not consider upstream or downstream effects that may cause damage to other parts of the basin.

Another concern is the behaviour of drainage structures under extreme events, especially structures designed many years ago when land use occupation was different and data was scarce. This is even more the case if – as happens in many places – conduits were combined for drainage of sewage and runoff (a practice no longer widely adopted). These have to be examined under the new criteria, and adapted or changed accordingly. This is why hands-on experience by professionals is crucial for the design of adequate storm runoff structures.

Decisions are not always 100% technical in aspect, even after detailed technical assessment. Accordingly, the most important decisions should be taken by practising engineers with many years of experience: only these experts can make qualified judgments, taking into account stakeholders' opinions and the need for political approval. Such factors cannot be modelled or simulated.

8.4.2 Information requirements to support decision-making processes

This section presents an overview of the information required by different groups, aside from water engineers, to support decision-making activities related to urban water management. One of the key problems is the fact that many urban communities are unable to see the connections with the natural environment in the way that their elders do. This is the result of emigration of people from their ancestral lands where environmental behaviour was orally transmitted from generation to generation, alongside intuitive knowledge on how to manage water resources and respond to threats of flooding.

Rainfall

The general public generally has a good perception of what is considered to be an 'average' or 'normal' rainfall event for their surroundings and different times of the year. Today, satellite information and the media and internet, permit people to learn more such topics, as well as information about conditions that might generate potential storms. For example, in 2005, the city of New Orleans in the United States received warnings for days prior to the arrival of hurricane Katrina, which devestated the city. As a result, most people left the area in time, but those that chose to stay did so at their own risk.

Technical measurements for these types of storms are evolving every day, and the instigation of preventative strategies are helping to mitigate losses. However, many communities in developing countries do not have easy access to the internet and the media is not prepared to provide the information necessary for people to take action and protect themselves.

There are two training activities required to remediate this specific issue. The first is the training of broadcasters, to enable them to provide comprehensible weather forecasts. The practice in developed countries and some developing countries (e.g Chile) provides examples as to how this may be accomplished. In this way, the forecast of events can trigger alert systems in flood-prone areas, and communities can be mobilized rapidly with the help of media broadcasts. However, people need to be taught how to interpret the information given out during broadcasts.

The second action is a more hands-on, time-consuming practice involving the general public. It involves teaching individuals in flood-prone areas to measure rainfall through the construction of homemade devices (calibrated with standard equipment), and asking them to record when the level of rainfall is above normal. Although such rainfall measurements are not scientifically exact, they may be used as indicators by residents to assess how much rain has been falling. In general, more than 25 mm in twenty-four hours is considered by some as the point where preparation for evacuation should begin and determines what action should be initiated.

The acquisition of rainfall data by households has several other benefits. The rainfall gauge network expands noticeably, enabling more information to be acquired quickly and cheaply, permitting better studies. In addition, households thereafter possess their own record of rainfall, which can function as a signal of possible damaging events heading their way. In such cases, they can take immediate action to protect themselves.

Another method of creating public awareness is the placement of flood marks at key locations in the local area. Older members of the community can be asked to identify the heights of previous flood events. The markers can be used as indicators to warn local people that a serious flood event is expected. For example, in 1999, mudslides in Venezuela were preceded by several signs hinting at what was to come. Some creeks had overflowed days before, small landslides had occurred in several places, and flows in several mountain creeks were unusually high for the time of year. However, people did not know how to interpret these signs and suffered the consequences. This combined information should have enabled the authorities to realize the imminent dangers.

Topography, land use and floodplain delineation

Elevation curves are difficult for the inexperienced eye to appreciate. Government officials, stakeholders from NGOs and non-engineering professionals therefore need to be instructed in this skill. The greater their knowledge concerning flood levels, the areas involved and other characteristics, the better equipped they will be to take appropriate decisions and actions regarding potential damage from extreme events. It is also important that they better understand the work of the professional responsible for floodplain delineation and the setting of evacuation routes, safety zones and the positioning of important infrastructure (such as emergency services). Geographic information systems (GIS) can be a powerful tool in this area if sufficiently detailed data is available for the different layers of information required to create a base map.

8.4.3 Understanding of hydraulic concepts

The practice among professionals is to employ technical terminology when communicating. However, when speaking with the layman, this may result in information not reaching the audience, or being fully understood. For this reason, a short guide describing the most usual concepts handled by decision-makers in drainage basins is useful (see UNESCO-IHP, 2003; WMO, 1998 www.projectwet.org for examples).

These guides should include pictures of what a certain magnitude of flow would look like in a known river or channel, and mention of what this magnitude represents in terms of danger to property or human lives. Such concrete terms are more useful than abstract numbers or concepts for the average layman. A short briefing or seminar could be held to highlight the importance of these concepts in decision-making. These seminars have to be given by experienced professionals who can present case studies (see below), preferably undertaken by themselves.

Probability, risk analysis criteria

Concepts of return periods, levels of damages in accordance with water levels, and the combined analysis of these should be encouraged to promote better understanding of the frequency of occurrence of certain extreme events, and also to understand the potential damage that can occur from these events.

Tucci and Bertoni (2003) refer to several workshops held in Latin America where recommendations were made on urban drainage management. In one held in Argentina at Rio Carcarana in 2002, the lack of education of professionals in municipalities was specified as part of the problem in urban drainage management (along with lack of information exchange, coordination between different levels of public administration, land-use planning, risk analysis and others). As can be seen, the training needs mentioned in this chapter are widespread and steps need to be taken urgently in this direction.

8.5 CAPACITY-BUILDING INITIATIVES

8.5.1 Human resource development

Managing water resources requires skilled people with a good knowledge of the different aspects of water. As stated above, there are not enough human resources to appropriately study all the urban drainage problems in every city in the tropical regions. For this reason, everyone involved in flood-prone areas should be aware of their effects.

Most of these tropical countries are developing countries. Data on scientific and research personnel ratios published in the first United Nations World Water Development

Report (WWAP, 2003) reveals a ratio of about 8,000 scientists and engineers per 100,000 inhabitants in developed countries. In developing countries, only 60 to 80 scientists or engineers are accounted for per 100,000 inhabitants. Therefore, governments in the developing world need to be aware that investment in science and research are and essential prerequisite to solving the many challenges they face.

8.5.2 Learning from past experience

During emergencies local stakeholders play an important role in the decision process. This process can be assisted by examining previous examples of extreme events that have occurred in the vicinity or in other similar places, and considering their impacts. This can also help promote an understanding of the importance of urban drainage studies, master plans and the design of protective infrastructure, working towards prevention by avoiding certain types of construction on flood-prone areas.

One of the best ways to learn about almost anything is by actually undertaking the required activity. However, in storm runoff studies and design this is not possible for two reasons: first, because extreme events occur infrequently and, second, because there is usually a lack of real-time monitoring of the different parameters involved.

Although, some cases have been fully documented (Japan, for instance, where basin networks are fully equipped with measuring instruments), the practitioner often finds him or herself designing by empiric methods with sometimes scarce data. This is why, whenever possible, records need to be made of every possible event, its hydrology (precipitation, flows, velocities), flooded areas, height of water level at the peak of the flood, and evaluation of damage. This will help increase understanding of the behaviour of a basin under certain rainfall events.

Case studies are of particular interest to engineers, helping to identify the necessary parameters that influence a design. In recent years an effort has been made to publish well-documented cases. A group of investigators from IPH/UFRGS (Instituto de Pesquisas Hidráulicas/Universidade Federal do Rio Grande do Sul) in Brazil (Porto Alegre) have been undertaking solid work in this area. The 1999 Venezuelan mudslides were documented in an UNDP publication that prompted further studies, which were performed by the 'Institute of Fluid Mechanics' (Instituto de Mecanica de Fluidos') of the Central University of Venezuela. These focused on causes, behaviour of mudflows, probable future occurrences, delineation of potential and severe damage areas, and recommendations for the rebuilding of infrastructure in the affected areas. Japan has also performed extensive studies on the relation between land-use changes and the characteristics of runoff responses, which help in flood control analysis.

For the practising engineer, there are an increasing number of publications where he or she can learn about the causes and effects of extreme events floods. These should assist with the design of necessary structures to protect lives and infrastructures, and also to signal preventive measures such as the abandonment of possible sites of frequent flooding where protection against extreme events is shown to be uneconomic.

On-site experience is probably the best method of learning. To undertake this, water agencies must have monitoring devices and sufficient staff (both professionals and technicians) to perform the job effectively and efficiently. In most developing tropical countries, emphasis is placed on water supply rather than drainage problems, which are handled reactively when they occur.

Education is therefore essential to achieving flood risk reductions and consequent benefits, such as lower reconstruction costs and no life of loss (either directly or indirectly as a result of water-induced sickness). Case studies may help by demonstrating these benefits to non-professional decision-makers. This should also help to ensure that funding for actual actions is allotted in good time, and not only when emergencies arise and recovery is needed, which is unfortunately often the case.

8.6 TRAINING CENTRES AND INSTITUTIONAL SUPPORT

8.6.1 Links with academic institutions for the development of tools

As mentioned previously, university chairs, as well as regional centres, may work together to train much-needed technical or professional staff. These institutes are also in the best position to develop tools to better study the hydrological cycle in their region. The creation of networks enables information exchange, the common study of cases, and solutions can be discussed via the internet or at occasional regional seminars.

At present, several university networks are already working on different areas. Urban drainage can easily be added as another topic for study and discussion within these networks. Efforts may also be combined with other organizations and NGOs that deal with water issues, for example, the Global Water Partnership, the Inter-American Dialogue on Water Management, and hydraulic engineering societies in different countries.

In general, academic institutions are the ideal centres for training the human resources needed, from the field technician to the specialized professional. They are also the ideal unit for research and development of newer and better tools to study drainage behaviour. Regional centres should help to foster programmes through adequate funding (a problem that many universities face in the developing world), and by providing, if required, professional expertise to teach some of the courses.

8.6.2 Role of the regional centres and international agencies

Considering the different stakeholder groups described above that need training, the amount of training required is significant. The organization of courses, seminars, workshops and other training activities should be primarily the responsibility of universities, but there is also a role for both governmental and non-governmental organizations to play with the support of international agencies such as UNESCO International Hydrological Programme (UNESCO-IHP), the World Bank and the Global Environmental Facility (GEF).

Organizations such as universities and government offices that deal with flood hazards need to work together (supported by international organizations such as WMO, UNESCO-IHP, UNDP and UNEP) to create programmes, seminars and hand out material that will disseminate the basic information required to start actions in effective urban drainage management. Provided these organizations collaborate to programme and fund a comprehensive and coordinated capacity-building programme, knowledge and skills can be spread more effectively and efficiently across developing countries in the humid tropics.

Professionals have to be hired, in-country if possible, to manage and present seminars. This is a difficult task given the limited number of experts in such countries.

Regional centres can assist here as many possess useful hydrologic data and, in some cases, data on flood measurements and damage registered during past events. They also have trained water professionals, who although they may not be experts in urban drainage, can acquire the necessary tools to teach on the subject. Furthermore, they have the necessary personnel to organize the logistics that these types of seminars require, as well as expertise in obtaining the necessary funds.

In 1975, UNESCO established a programme to promote water-related issues. The International Hydrologic Programme (IHP) was created to prepare a scientific study of the hydrological cycle and formulate strategies and policies for the sustainable management of water resources. IHP is represented in about 160 countries, and works closely with other organizations such as the World Bank, WMO, GEF and private donors.

UNESCO-IHP works closely with at least three centres in the tropics: the Institute for Hydraulic Research of the Federal University of Rio Grande do Sul (IPH/UFRGS), in Porto Alegre, Brazil; the Centre for the Humid Tropics of Latin America and the Caribbean (CATHALAC) located in Panama; and the Humid Tropics Hydrology Centre located in Kuala Lumpur, Malaysia. Another centre for urban water management for Latin America and the Caribbean has been created recently in Colombia, and a further centre in Africa has yet to be established.

These centres, work closely with governments and other institutions (such as universities) and play an important role in promoting seminars, courses and the educational requirements mentioned in this chapter. Because of their international status as UN-backed programmes, they can coordinate and foster collaboration between different actors within a country. This is very important in places where stakeholders' interests might interfere with solutions of general interest to the public. Ideal organizations for the management water resources, such as basin agencies, can be established with back-up from such centres. This form of collaboration concerns not only human or material resources, or fundraising (multinational agencies, bilateral, foundations), but the key activity of educating and preparing stakeholders in sustainable water management practice.

Furthermore, because UNESCO-IHP possesses a database of professionals with experience in this technical field, instructors can be found from different regions to provide different points of view on specific cases – an enriching activity that encourages technological transfers to less-developed nations. UNESCO also has a network of university chairs which deal with water issues. These work closely with UNESCO-IHP and can prove very useful for training in-country professionals within the regions.

8.7 CONCLUSIONS

Training of much-needed professionals to manage urban drainage problems in the tropics is an urgent necessity, as cities are growing at an increasing rate and land occupation occurs with little or no control. The measurement of hydrologic data is essential to understanding flood phenomena. Therefore hydrologic networks must be increased, and the technicians required to handle the equipment must be trained. In parallel to this effort, engineering programmes should include more preparation on drainage problems, floods and their control.

In general, it is important for people to learn about their surroundings from a flood-hazard viewpoint, and in particular, government officials involved in planning and decision-making. The more people know about the environment that surrounds them, the greater their capacity to react in emergencies. Urban land-use planning is essential for prevention measures, as is the enforcement of regulations. (In many countries flood-prone areas have been separated, only to be occupied by poor people who have nowhere to go.)

Finally, urban drainage forms part of a holistic water resources management process, of which the natural element is the basin. As we increase our knowledge of how a basin works under different hydrologic conditions, we will be able to better prevent casualties from flooding, and design better infrastructures to capture stormwaters.

REFERENCES

Gunston, H. 1998. *Field Hydrology in Tropical Countries: A Practical Introduction*. Rugby, UK: Practical Action Publishing.

Ligget, J. and Ettema, R. 2001. Civil Engineering Education: Alternative Paths. *Journal of Hydraulic Engineering*. ASCE. Vol. 127, No. 12, December 2001, pp. 1041–51.

Mendiondo, E.M. 2005. *An Overview on Urban Flood Risk Management*. Escola de Engenharia de Sao Carlos, Univ. de Sap Paulo: Minerva, 2(2): 131–43.
http://www.fipai.org.br/Minerva%2002(02)%2003.pdf (accessed 20 December 2009).

Munich Re. 2001. *Topics, Annual Review: Natural Catastrophes 2000*. Munich.

Tucci, C.E.M. and Bertoni, J.C. 2003. *Inundações Urbanas na América do Sul. WMO, GWP, ABRH*. http://www.iph.ufrgs.br/corpodocente/tucci/DisciplinaDrenagem.pdf (accessed 20 December 2009).

WMO. 1996. *Guide to Meteorological Instruments and Methods of Observation*. http://www.wmo.int/pages/prog/www/IMOP/publications/CIMO-Guide/CIMO%20Guide%207th%20Edition,%202008/CIMO_Guide-7th_Edition-2008.pdf.

WMO. 1998. *Hydrology of disasters*. Proceedings of the technical conference held in Geneva. World Meteorological Organization (WMO).

WWAP. 2003. *United Nations World Water Development Report*. Paris: UNESCO Publishing/ Berghahn Books. http://www.unesco.org/water/wwap/wwdr (accessed 20 December 2009).

Sources of further information

ARMCO. 1950/1971/1981. *Handbook of Culvert and Drainage Practice*. (Spanish). American Iron & Steel Institute.

ASCE. 2001. *Journal of Hydraulic Engineering*. Special issue: Teaching hydraulic design. Vol. 127, No. 12.

Asociacion Iberoamericana de instituciones de ensenanza de la ingenieria. 2006. El profesor de ingenieria, profesional de ingenieros en iberoamerica. IV encuentro iberoamericano de instituciones de ensenanza de la ingenieria (presentacion).

AWRA. 1991. *Urban Hydrology* (Symposium proceedings). Technical Publication Series TPS-91-4. Maryland, US: American Water Resources Association.

Bolinaga, J. 1979. *Drenaje Urbano*. Instituto nacional de Obras Sanitarias.

Franceschi, L. 1984. *Drenaje vial*. Fundacion Juan J. Aguerrevere.

HERON. 2004. *Special Publication on Risk Management*. Several authors. 49, No. 1.

Hidalgo, L. and Hidalgo, J. 2006. *El arte de medir la lluvia tropical*. Caracas: Fundacion Polar.

Maksimović, C., Saegrov, S., Milina, J. and Thorolfsson, S. (eds) (2000) *Urban Drainage in Specific Climates. Vol. II: Urban Drainage in Cold Climates*. Paris: UNESCO–IHP.

MIT. 2000. *CEE New Millennium Colloquium*. Department of Civil and Environmental Engineering.

Ochoa-Iturbe, J. 2006. El Ingeniero del siglo XXI. Conference for the 3rd Interamerican Congress of Civil Engineering Students. Margarita Island, Venezuela.

Organización de los Estados Americanos (OEA). 2001. Estrategia interamericana para la promoción de la partipación pública en la toma de decisiones sobre Desarrollo Sostenible. Washington DC.

Primer Congreso Venezolano de Ensenanza de la Ingenieria. 2006. Conclusiones y recomendaciones.

Sheaffer, K., Wright, K.R., Taggart, W.C. and Wright, R. 1982. *Urban Storm Drainage Management*. New York: Marcel Dekker.

Salazar, F. and Pornes, C. 2003. *La ingenieria del siglo XXI, para una sociedad sostenible*. Guatemala: Universidad Rafael Landivar.

Yackovlev, V. 2006. *El ingeniero del 2025*. Caracas, Venezuela: Academia Nacional de Ingenieria y el habitat.

Index

T - #0616 - 071024 - CO - 246/174/8 - PB - 9780415453530 - Gloss Lamination